Sky Blue Stone

THE CALIFORNIA WORLD HISTORY LIBRARY

Edited by Edmund Burke III, Kenneth Pomeranz, and Patricia Seed

Sky Blue Stone

The Turquoise Trade in World History

Arash Khazeni

UNIVERSITY OF CALIFORNIA PRESS

Berkeley · *Los Angeles* · *London*

University of California Press, one of the most
distinguished university presses in the United States,
enriches lives around the world by advancing scholarship
in the humanities, social sciences, and natural sciences. Its
activities are supported by the UC Press Foundation and
by philanthropic contributions from individuals and
institutions. For more information, visit www.ucpress.edu.

University of California Press
Berkeley and Los Angeles, California

University of California Press, Ltd.
London, England

Library of Congress Cataloging-in-Publication Data

Khazeni, Arash.
  Sky blue stone : the turquoise trade in world history /
Arash Khazeni.
    pages   cm. — (The California world history
library ; 20)
  Includes bibliographical references and index.
  ISBN 978-0-520-27907-0 (cloth : alk. paper)
  ISBN 978-0-520-28255-1 (pbk. : alk. paper)
  ISBN 978-0-520-95835-7 (ebook)
    1. Turquoise—History.   2. Mineral industry—
History.   3. Mines and mineral resources—History.
I. Title. II. Title: Turquoise trade in world history.
  QE394.T8K53   2014
  381′4287—dc23   2013041585

Manufactured in the United States of America
23 22 21 20 19 18 17 16 15 14
10 9 8 7 6 5 4 3 2 1

For Layla and Aiden

راستی خاتم فیروزه بواسحاقی
خوش درخشید ولی دولت مستعجل بود

In truth the turquoise ring of Abu Ishaq
Flashed finely but then faded away

—Hafiz

Contents

Illustrations

PLATES

following p. 92

Preface

Looking south from the portal of the bazaar on the central square in the city of Isfahan, you can see the turquoise colored dome and minarets of the royal mosque reaching from earth to sky. The square, known as the Pattern of the World (Maydan-i Naqsh-i Jahan), was built in the early seventeenth century in the reign of the Safavid monarch Shah ʿAbbas I as the crowning monumental space of his capital city. When I first stepped into the Maydan, Isfahan appeared to be a turquoise city, but little did I know what that meant or that the city had once been a major hub of the global turquoise trade. It was 2001, and I was a graduate student doing research for my thesis on imperial encounters with "tribes" in nineteenth-century Iran—the history of turquoise could not have been further from my mind. Later on, in the course of reading Persian travel books from the nineteenth century, I came across a rather detailed description of the turquoise mines of Nishapur in Muhammad Hasan Khan Saniʿ al-Dawla Iʿtimad al-Saltana's 1882 text *Matlaʿ al-Shams* (Land of the rising sun), which chronicled the Qajar Dynasty's efforts to reclaim the mines and gain control over their output and trade. After digging below the surface more and finding a practically untouched genre, Persian mineralogical texts on precious stones, produced in Islamic courts between the fifteenth and nineteenth centuries, I realized that there might be something there to pursue further.

Writing the history of this stone, however, posed certain historiographical dilemmas. To begin, very little existed in the way of quantitative source material on the turquoise trade—no substantial East India Company records or merchant correspondence—apart from fleeting references to the stone in lists of commodities. This dearth of quantitative and commercial documentation was related to broader questions of perspective and narrative. With time, I began to see the oddities of the history of turquoise, which did not fit the patterns of existing narratives on commodity chains. It was not the story of a colonial commodity—such as sugar, cotton, gold, or diamonds—whose exploitation enslaved and oppressed indigenous populations on the periphery of a global capitalist order with Europe at its center. Neither was turquoise a globalized commodity—on the scale of silks and spices—widely in demand among European consumers or that brought great wealth to East India Company merchants as a staple of commerce between Asia and Europe. Thus, in writing the history of the turquoise trade I had to come to terms with the absence of quantifiable data and the sheer fact that turquoise was never a major commodity in the European commercial economy. Its market was not in Europe; it was coveted elsewhere. The stone was an object not of westward oceanic trade but of the land-based caravan trade in the other direction. This was a less familiar tale, one that offered a break from prevailing, bullion-oriented accounts of the economies of early modern Eurasian empires.

The Eurasian turquoise trade was not less documented; it was differently documented. Instead of foreign company records, the sources for its study are an archive of Persian mineralogical literature that conveys the cultural meanings attached to the trade and exchange of precious stones. While turquoise was rarely more than an exotic and a curiosity from the vantage of the commercial economy of Europe, it was a hallowed natural object and color in the tributary economies of Islamic empires that moved between the steppe and the sown in early modern Eurasia. Across Timurid Central Asia, Safavid Iran, and Mughal India, where natural histories in the genre of "books of stones" were produced, turquoise became an object of imperial interactions and exchanges, traded, gifted, and looted by Islamic dynasties and their subjects. Its sky-blue color came to adorn the vaulted azure-tiled monuments of oasis cities from Samarqand to Tabriz to Sindh.

Still, as I write these words, I can anticipate some of the critiques that will be sent my way. Persianate books of stones will be deemed formulaic and of a fixed stock—but then again, the same could be said for any corpus of source materials, including European company correspondence, accounts, bills, and receipts. Some will likely dissent that I have not done enough to ground the turquoise trade in the religions of Islam or Safavid Shi'ism, but such an essentialist project was never the purpose of this book, which attempts to trace the material culture of a stone, nor is it substantiated by the empirical materials, which incline instead toward knowledge of the natural world. Still others may ask what is environmental about this tale apart from the fact that turquoise comes out of the ground, yet given that most histories of Eurasian commodities have been more concerned with commerce, politics, and religion than actual material culture (and the vernacular sources for the study of substances), this subtlety alone could be enough. Environmental history and natural history are perspectives and may take on different narratives when viewed from other parts of the world. This book sets the material culture of the turquoise trade, one tale of the quest for the earth's mineral resources, against the history of pastoral societies moving between the steppe and the sown to carve out Islamic empires and build oasis cities across the Eurasian expanse.

Writing this book, I was fortunate to find help from various scholars, boon companions, and friends, and I give thanks to them here. For ideas, inspiration, and intellectual kinship, my gratitude begins with my adviser Abbas Amanat, who taught me what I know of the craft of Safavid and Qajar history (although I still have much more to learn) and supported this project from its beginning. The creative influence of Nile Green and his perceptive and pathbreaking traverses across the fields of South Asian, Central Asian, and Near Eastern history have been an inspiration to me, and I owe much to his deep insights. Since I moved to Los Angeles, the subtle mentoring and generous camaraderie of Sanjay Subrahmanyam and his pioneering trail of work into the tangled histories of the Indo-Persian world have sparked the direction of my research.

At various stages, Sebouh Aslanian, Terry Burke, Naindeep Chann, Rudi Matthee, Alan Mikhail, Farzin Vejdani, and Waleed Ziad offered frank and critical advice that made this a better book.

In the History Department at Pomona College, I have been fortunate to find a thriving intellectual abode, and I am grateful to friends and colleagues there: Angelina Chin, Pey-Yi Chu, Gary Kates, Sidney Lemelle, April Mayes, Victor Silverman, Tomás Summers Sandoval, Miguel Tinker Salas, Helena Wall, Ken Wolf, and Sam Yamashita. Helena and Victor read all of the manuscript with care, and I hope they will see the influence of their critiques here. Char Miller gave me a primer and a reading list in urban environmental history that I am still working my through. I also thank Gina Brown-Pettay, the heart and soul of the History Department. Other friends and colleagues at Pomona who supported and encouraged me: Tahir Andrabi, Dru Gladney, George Gorse, Kathleen Howe, Zayn Kassam, Ben Keim, Jade Star Lackey, Pardis Mahdavi, David Oxtoby, Virginie Pouzet-Duzer, Cynthia Selassie, Shahriar Shahariari, and Jonathan Wright. The Pomona students who took an interest in and helped this project: Clare Anderson, Anisha Bhat, Camille Cole, Elizabeth Kokemoor, Aaran Patel, and Leyth Swidan. I also thank Pomona College for the generous institutional support that I received while researching and writing this book, including the Downing Fellowship, which allowed me to be a visiting scholar at the University of Cambridge and explore the Persian books of precious stones in the Edward G. Browne Collection.

I am grateful to the people who helped me in the course of my travels and research. Without them, this book would not have been possible. In London: Assef Ashraf, Khodadad Rezakhani, and Thomas Wide. At Cambridge University: Charles Melville, Paul Millet, and David Pratt. In Tehran: my friend and colleague Mohammad Reza Tahmasbpour, along with Zahra Asadiyan and Akram Ali Bayayi of the Golestan Museum and Photographic Archives, graciously assisted in locating rare nineteenth-century photographs relating to the turquoise trade. Closer to home, Tofigh Heidarzadeh at the Huntington Library guided me through the Islamic history of science as well as on strolls through the library's intricate botanical gardens. At the Claremont Colleges' Honnold Library, Carrie Marsh and Lisa Crane provided invaluable access to a rare collection of early modern texts on mining and metallurgy. I am also grateful to my dear friend the one and only Gillian Schwartz, for the hand-drawn maps she contributed to the book.

The decision to write a book on turquoise and its trade came after a discussion with Jennifer Wapner, then the natural history editor of University of California Press, at the annual conference of the American Society for Environmental History in Portland, Oregon, in 2010. That meeting, as well as subsequent conversations with Niels Hooper, convinced me that turquoise was a story worth telling, and that is how this book began. I am grateful to Jennifer and Niels for taking a chance on this project, even back when, admittedly, there was not much there, and to the whole editorial team at University of California Press, including Chalon Emmons and Kim Hogeland, for seeing this book through to press. Juliana Froggatt meticulously edited the manuscript.

Most of all, I thank my family. In Tehran and on road trips all across Iran, Faizeh was my dear partner of the highways, cities, and mountains and showed me endless hospitality and companionship. My mother Farah supported this project from the start and inspired me with her own recollections and stories of the sky-blue stones of Nishapur. Dana, as always, was my closest friend through it all; she sustained me in more ways than I can easily say. This book would not have been possible without the inspiration and love of Layla and Aiden, with the turquoise eyes, and I dedicate it to them with the deepest affection and promises of further journeys to come.

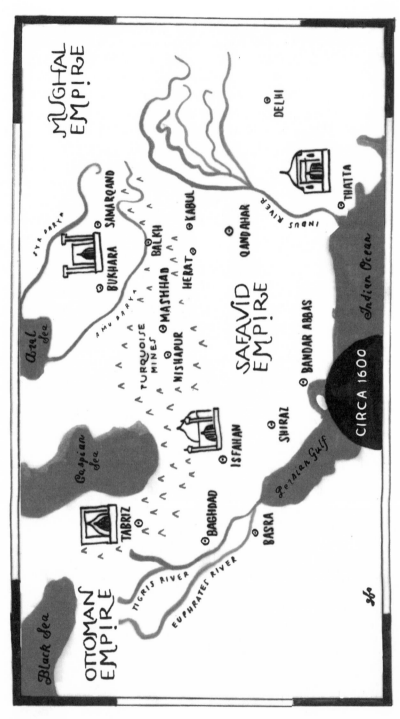

MAP 1. Eurasian empires: the turquoise mines of Nishapur and the blue cities of the eastern Islamic world in the age of the Safavids, Mughals, and Ottomans. Map by Gillian Schwartz.

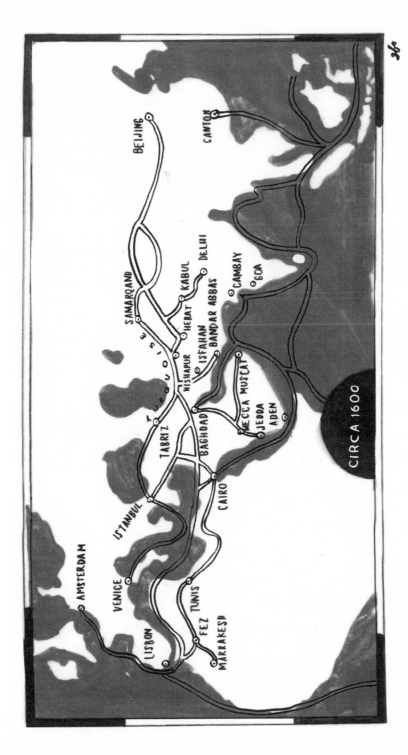

MAP 2. Eurasian crossroads: long-distance trade routes spanning early modern Eurasia by land and sea. Turquoise spread from its remote point of origin, mines in eastern Iran, across the land-based caravan routes connecting Central Asia, South Asia, and the Near East. Map by Gillian Schwartz.

Introduction

The Turquoise Ring of the Emperor Jahangir

In the spring of 1613, the Mughal emperor Jahangir dispatched Muhammad Husayn Khan Chelebi, a merchant of gems and other precious objects, to Safavid Iran with letters of introduction and orders to purchase rarities for the royal estate in India. Chelebi met Shah ʿAbbas I in the eastern Iranian province of Khurasan and presented him with a letter from the Mughal emperor. Most important among the list of items that the merchant was charged to find was quality turquoise, but the Safavid monarch informed him that the precious stone was under royal monopoly and could only be gifted by the shah himself. Following this pronouncement, Shah ʿAbbas chose one of his personal attendants to turn over six bags of turquoise, containing thirty seers (roughly 675 grams, or about 1.5 pounds) of ore, to the merchant from India. The shah included with the gift a letter to Jahangir professing his brotherhood and friendship while apologizing for the inferiority of the turquoise and reporting that the gem was no longer mined as it once had been. On receiving the bags of turquoise, Jahangir lamented that the quality was indeed poor, writing in his memoirs, "No matter how hard the gem carvers and setters tried, they couldn't find a stone worthy of being made into a ring."[1]

The gift of turquoise from the Safavid shah to the emperor of Mughal India in 1613 was embedded in the tributary imperial networks of early modern Eurasia. Among the post-Timurid Islamic empires of the Near East and South Asia—the Safavids, the Mughals, and the Ottomans—

sky-blue turquoise from Iran circulated widely and became valued as an object of imperial tribute and exchange. It was fitting that Jahangir, whose insatiable taste for jewels was legendary, should seek turquoise from the mines of the Safavid Empire. But his search for the wondrous blue stones of Iran was entwined in the ongoing imperial contention between the Safavids and the Mughals over the Afghan city of Qandahar, an outpost between their empires. Qandahar had been promised to the Safavid shah Tahmasp I by the Mughal emperor Humayun, who had been driven from the Indian subcontinent by the Afghans in 1540 and had taken refuge in Iran before reclaiming his empire in India with the aid of Safavid troops. But into the first decades of the seventeenth century, as Jahangir's merchants set out to find Persian turquoise, the Mughals had yet to relinquish the city. The poor quality of the turquoise that Shah 'Abbas sent, when he was merely a few miles from the city of Nishapur and the most esteemed mines of turquoise in the world, acquires fuller meaning when viewed in the context of this struggle over the Afghan crossroads between the empires. With the question of Qandahar unsettled, it may well have been that Shah 'Abbas withheld from Jahangir the choice turquoise from the old mines outside Nishapur. Or it may have been that the famed old turquoise mines of gem-quality stones were becoming depleted in the seventeenth century, as the shah seems to suggest in his letter.

A miniature painting that Jahangir commissioned in 1618, representing his dominion over the Indian subcontinent and Central Asia, evokes this gift of turquoise and the unsettled nature of the Afghan city of Qandahar. Drawn by the Mughal court artist Abu'l Hasan, the illustration depicts Jahangir and Shah 'Abbas embracing while standing on a globe of the world, set against the backdrop of a celestial turquoise sky (see plate 2). A larger-than-life Jahangir balances on a sleeping lion that straddles India and Central Asia as a diminutive Shah 'Abbas stands on a delicate lamb being pushed out of Asia. The "world-seizing" Jahangir appears with a ring of sky-blue turquoise stone on his right hand. To find turquoise entangled in the imperial history of early modern Eurasia is fitting, for, as the following pages argue, the stone was an object of imperial interaction, and the early modern turquoise trade flourished through the emergence of Islamic tributary empires of pastoral nomadic origins that moved from the tent to the throne to build imperial cities and become integrated into global routes of trade and travel linking the Near East, Central Asia, and South Asia.[2] Turquoise became an imperial stone and color in the tributary economies and material culture of early modern Islamic empires, which negotiated their power with rival states and their own subjects through the exchange and display of regalia and nature's objects. From the lands of the interconnected Timurid, Safavid, Mughal, and Ottoman Empires, where turquoise appeared as an object of imperial power projected in vivid color displays, the stone and its culture traveled across the world.

This book traces the journeys of turquoise from its remote point of origin outside the city of Nishapur in eastern Iran across the Near East, India, Central Asia, Europe, and, in the end, the Americas. An opaque sky-blue phosphate of aluminum and copper formed by nature in rocks below the surface of the earth, "turquoise" became known as a mineral substance in early modern networks of travel and trade. Found exclusively in desert environments, its deposits were historically mined in a wide mineral-bearing stratum extending from Egypt through Iran to Tibet, with the most precious stones unearthed from the mines of Nishapur. Turquoise evolved into an object of imperial interaction and exchange among the empires of early modern Islamic Eurasia. By the sixteenth century, as it traveled from Nishapur through the blue-tiled cities of the eastern Islamic world and farther, to Venice, Paris, and other European markets, it was coveted as a strange and exotic object from the East. Becoming associated with the Turks and the trade routes that carried the gem across the Ottoman Empire to Europe, the stone was called *pietre turchese* in Italian and in French *pierre turquoise,* or "Turkish stone."

As turquoise traveled, something of its meaning went with it. Turquoise was among the tinted stones from Asia that gave substance to the perception of the color blue. Brilliant natural substances—turquoise, cobalt, and lapis lazuli from Iran and Afghanistan—were traded across the early modern world and catalyzed the creation of the previously uncommon color of blue as a cultural phenomenon. Turquoise was the color of the sky, and it was color that brought turquoise into demand and defined its culture as an object. Where turquoise reached, from the tents of Central Asian pastoral nomads to the royal courts of Eurasian princes, it left its meanings behind, prized for the nature of its celestial blue. Turquoise was worn as an ornament and as a jewel adorning rings, cameos, and amulets; dusted the leather bindings of books; was inlaid on the surface of shields, bridles, and weapons of war; ground into powder, was taken as medicine; and regarded as one of the seven colors (*haft rang*) of heaven, was adopted for the palette of tiles fired in the workshops of ceramicists and appeared in Islamic Eurasia as the color of imperial cities and their architectural monuments. The

Eurasian turquoise trade flourished throughout the early eighteenth century, until the fall of the Safavid Empire in 1722 and the subsequent ruin of the old mines of Nishapur led to its ebb. By the nineteenth century, when the Qajar dynasty attempted to revive the mines, colonial empires had eclipsed the tributary empires of Islamic Eurasia, and the imperial meaning of the turquoise trade faded away and was lost. In the 1890s, the reopening of lost Aztec mines in the Americas, along what came to be known as the turquoise trail, unearthed more accessible sources of the stone that rivaled the Persian blue.

EMPIRES AND ENVIRONMENTS

Through an examination of turquoise and its trade, this book attempts to locate the history of Islamic Eurasia in the context of world environmental processes and global encounters between empire and nature. Existing literature on the subject has considered the ways that expanding European empires—and their naturalist and scientific knowledge systems—discovered, classified, and transformed environments at the ends of the earth in efforts to control local natural resources since the early modern period. Such groundbreaking works of global environmental history as *Ecological Imperialism: The Biological Expansion of Europe, 900–1900* by Alfred Crosby, *Green Imperialism: Colonial Expansion, Tropical Island Edens and the Origins of Environmentalism, 1600–1860* by Richard Grove, and *The Unending Frontier: An Environmental History of the Early Modern World* by John Richards have examined these encounters, contending that knowledge of nature and environments, acquired through the sciences of natural history, botany, and geology, gave European empires power over the resources of the earth.[3] A closely related strand of historiography has examined European encounters with nature from the vantage of the history of science and the making of European natural sciences. Prime areas of focus in this literature are the early modern European fascination with collecting nature, plants in particular, and the sciences and natural histories that emerged to describe and classify exotic objects and commodities—the "marvelous possessions" and "worldly goods" exchanged around the world in an age of European exploration, commerce, and renaissance.[4] The trade in nature—the quest to possess and know the resources of the earth—arose, it is suggested, through European voyages of discovery in the Indies.

In these works, the emphasis has been on European maritime empires and their voyages of exploration, global trade, and scientific discoveries. This conventional view of global trade and exploration as the outcome solely of European voyages of discovery has already begun to fray, as seen, for instance, in the innovative work of Kapil Raj, which insists that Europe's knowledge of nature was the product of encounters with indigenous knowledge systems from across the world.[5] Departing from the implication that natural history was directly accessible to European explorers in Asia, Raj examines the role of intermediaries and indigenous or vernacular practices and knowledge systems in European constructions of nature in the Indies. Far from the notion that modern science was created by Europe and diffused around the world, Raj convincingly details how sciences were made through intercultural encounters, the actual site of the exchange of goods and knowledge of their meanings. Yet despite this historiographical shift, the emphasis in the field remains on how indigenous knowledge entered and became embedded in European colonial sciences, and the story is still told indirectly, through European sources that have "translated" and retrieved vernacular knowledge of the natural world. Whether dealing with the New World, where presumably few vernacular written sources exist, or Asia, where Asian-language materials are abundant, these works tap into local knowledge systems only indirectly, through readings of European texts, and ultimately still emphasize the European imperial collection of nature. What seems to be missing in the literature is an effort to trace indigenous knowledge of the natural world at least in part through local, vernacular source materials.

The history of the Eurasian turquoise trade presented here parts ways with narratives of discovery and exploration of the Orient and the Indies by European trading companies and maritime commercial empires seeking commodities and other valued goods. The following pages turn instead to the land-based tributary empires that moved between the steppe and the sown across Islamic Eurasia and their encounters with—and quest to lay claim to—the natural world and its resources. *Tributary empires* refers to the vast, multiethnic land empires of the premodern and early modern world. Such regimes were land-based imperial spheres that ruled through the collection of tributes and revenues from subjects and subject kingdoms absorbed within the empire. These empires of difference were neither commercial nor colonial but rather based on an ongoing negotiation of power and a layered sovereignty between the empire and its satellites and subordinates.[6] The tributary empires of early modern Islamic Eurasia—Timurid, Safavid, Mughal, and Ottoman—were expansive world empires of pastoral foundations that included agrarian lands and oasis cities and sought dominion over production and trade. By their very nature, such empires mediated their authority with their subjects and conducted their diplomacy through the exchange of tributary goods and objects. Elaborate customs of display and the representation of imperial power created a culture of exchange among the post-Timurid Eurasian empires. Turquoise thus came to be immersed in the exercise and projection of royal power in a particular cultural and imperial world connecting the Near East, Central Asia, and South Asia. As seen through turquoise and its trade, Eurasia circa 1500–1850, far from being a colonial periphery, is reconfigured as a dynamic expanse: the setting of tributary Islamic empires that were a point of origin in the export of natural objects and their meanings around the globe.

The circulation of the natural substances of the earth has rarely figured as a subject of environmental history, a field still dominated almost exclusively by narratives of degradation and conservation, rooted in the literature on the American West, where the field originated. Histories of commodities and objects have rarely ventured to adopt an environmental perspective, while environmental histories have yet to take natural substances as a frame of analysis. The environmental perspective pursued here veers toward natural history and concerns the economy and the culture of turquoise, its origins, its uses, and its trade as a mineral substance.[7] The history of the Eurasian turquoise trade in the early modern and colonial periods brings the literature on the global circulation of goods and commodities down to earth, detailing how the sky-blue stones came to be trafficked and treasured as objects of exchange among tributary Islamic empires moving between the steppe and the sown.[8]

OBJECTS OF WORLD HISTORY

The production, circulation, and consumption of commodities in world history is well-explored terrain, yielding an array of studies on silks, spices, sugar, coffee, salt, and other global substances. Since the publication of Sidney Mintz's classic *Sweetness and Power: The Place of Sugar in Modern History*—a Marxist critique of European imperial discovery and conquest and the expansion of the capitalist economy over the natural resources of the colonized periphery—the historiography of commodity chains has taken a cultural turn, in works on such subjects as pineapples, plumes, and porcelain.[9] The study of Eurasian commodities has also seen the publication of a number of groundbreaking books predominantly focused on the trade of silk, including Thomas Allsen's *Commodity and Exchange in the Mongol Empire: A Cultural History of Islamic Textiles* and Rudolph P. Matthee's *The Politics of Trade in Safavid Iran: Silk for Silver, 1600–1730.*[10] More recent works have opened the discussion to include a broader range of goods and commodities, including spices, plants, tea, coffee, ceramics, and jewels.[11]

The story of the Eurasian turquoise trade adds more to this literature than simply the tale of another networked object. Works on the economic history of early modern Eurasian empires, including the fields of Mediterranean and Indian Ocean studies, have focused for the most part on European commercial capitalism and the material objects that Europe valued and profited from as commodities—silks, spices, and textiles—all set in motion through precious metal flows and globalized merchant diasporas plying the channels of the early modern world economy. The pervasive view is one of merchants exchanging goods and ideas across cultures and long distances in ways that closely parallel the ideology of globalization and seemingly reveal all parts of the world as naturally interconnected by flows of capital.

The Eurasian turquoise trade gets us away from these prevailing views, unearthing a vernacular economy and culture of inter-Asian exchanges based in the tributary networks of early modern Eurasian empires.[12] The journeys of turquoise reveal the history of a global object that did not matter to Europe but was greatly esteemed and took on profound meaning across the tributary Islamic empires of early modern Eurasia. The turquoise trade, like that of the cowrie shell, conveys the limits of Europe's familiarity with Asian economies and material culture and details vernacular economic networks and exchanges that Europe and its East India trading companies could not absorb.

At the same time, the turquoise trade elicits a consideration of different contexts and spaces of encounter and exchange in world history circa 1500–1800. The tale of turquoise, its material culture, and its trade shifts the perspective from European commercial empires and littoral ports to the Eurasian tributary empires of pastoral origins that moved from the tent to the throne to create a synthesis of the steppe and the sown and to build Islamic dynasties and oasis cities across Central Asia, Iran, and India between the early modern period and the nineteenth century.[13] It questions the conventional view that silk was the only exportable commodity from Iran before the opium and tobacco production of the nineteenth century: the prevailing account of foreign trade in early modern Iran has focused on silk and silver, in ways that echo the famous statement by the seventeenth-century French cleric Raphaël du Mans that "Persia is like a big caravanserai which has only two doors."[14] The Eurasian turquoise trade out of Safavid Iran shifts the perspective from silk roads to steppe roads and from commercial interactions between Asia and Europe to an exploration of the material culture of inter-Asian encounters and trade.[15] Instead of being analyzed for what it lacked—seemingly almost everything under the sun apart from silk, which was shipped to Europe in return for much-needed silver—early modern Iran, with its turquoise mines of Nishapur, emerges as a hub of inter-Asian tributary cultures and networks of exchange.[16] After more than five decades, Martin Dickson's call for closer concern with the "internal history" and "indigenous" views of the Safavid Empire remains as timely as ever.[17] In turning us from the westward-bound oceanic trade networks of European East India companies and maritime empires to the land-based, inter-Asian caravan trade of Islamic empires in the other direction, the history of the turquoise trade suggests another side of the economy of Safavid Iran.

EURASIAN CROSSROADS

The following pages adopt a Eurasian geographical frame by bringing together regions—the Near East, Central Asia, and South Asia—that are often taken to be distinct and separate "civilizations." The project of conceiving a cross-regional Eurasian history is by no means new. To take a well-known point of departure, Fernand Braudel's *The Mediterranean and the Mediterranean World in the Age of Philip II* rendered the sea itself a world-historical space over what he termed the *longue durée*, an almost motionless history linking Europe, Asia, and Africa.[18] In the field of Islamic history, Marshall Hodgson painted such a broad geographical canvas in *The Venture of Islam*, a work published in three volumes in 1974. Hodgson introduced the idea of "the Islamicate world," to cover what he perceived as the broad, intercommunicating, and hemispheric zone spanning Africa and Asia, from roughly the Nile to the Oxus (Amu Darya), the lands of Islam.[19] Turning to those lands, one finds a long tradition of seeing the span of the world in Islamic historical and geographical thought, such as the notion of the "seven climes," or *haft iqlim* (taken from the Greek *klimata*), of the inhabited parts of the world, present in the works of classical Muslim authors and the title of a Mughal-era Indo-Persian world geography by Amin Ahmad Razi.[20]

Among the most promising blueprints for approaching the broad expanse of Eurasian history appeared in the work of Joseph Fletcher, the pioneering scholar of the Sufi networks and steppe nomads of Central Asia. Fletcher's notion of "integrated history" emphasizes parallels, interconnections, and continuities across Eurasia and is concerned with "the search for a description of interrelated historical phenomenon" that linked societies throughout the region.[21] More recently, Sanjay Subrahmanyam presented a framework for reconfiguring the history of South Asia into a broader Eurasian space of conjunctures in two companion volumes titled *Explorations in Connected History* and a programmatic essay in the journal *Modern Asian Studies*.[22] This framework stresses "openness to other histories," an approach to early modern Eurasia through "the idea of crossroads," and a focus on contacts and encounters.[23] It diverges from comparative history by opting for interlinked and entangled histories while accounting for the fact that encounters and interactions could often lead to disconnections and constructions of difference. The framework of integrated or connected history also marks a Eurasian turn in world history, suggesting that the Asian expanse occupies a connective position like that which Braudel attributed to the Mediterranean Sea.

Histories of early modern and nineteenth-century Asia and the Near East have thus far been framed largely in terms of encounters with the West, but key questions still remain to be considered in regard to recovering and retracing the patterns of inter-Asian imperial encounters and exchanges. The history of the Eurasian turquoise trade illustrates those exchanges, detailing the Central Asian and Persianate connections that linked early modern Islamic empires.[24] Following the trail of turquoise and its material culture across the oasis cities of the eastern Islamic world, this book seeks to present a history of Asia from the inside out. The cities and pastoral steppes of the Eurasian expanse, far from being marginal hinterlands and frontiers, were a crossroads of imperial interactions, of economic and cultural exchange. With regard to the Safavid Empire, this focus also offers some distance from the old colonial emphasis on Iran's conversion to Shi'ism and subsequent separation from and role as "barrier of heterodoxy" against the surrounding Sunni empires of India and Central Asia.[25]

ON THE ECONOMY OF STONES

This study of the Eurasian turquoise trade draws on the previously unexplored genre of Persianate naturalist literature known as *javahirnama,*

"books of precious stones," as well as printed travel accounts and natural histories in Persian and European languages. *Javahirnama* was an
imperial genre of scientific literature about precious stones and the mineral substances of the earth, detailing their origins, uses, and trade. Such
books were part of the imperial economy of early modern Islamic Eurasia, written to inform sultans and shahs of the worldly objects and jewels
amassed in their treasuries following wars of conquest and the taking of
tributes. This literature is traceable back to a rich tradition of medieval
Arabic and Persian natural histories of precious stones, minerals, and
metals known as the *ahjar al-karima,* "great stones."

Manuscripts of these books of stones exist in archival collections in
Iran and the United Kingdom. The first Persian precious stone books
that I consulted are in the Majlis (or Parliament) Library in the Bahari-
stan district of Tehran. This collection, dating back to the Iranian Con-
stitutional Revolution of 1906–11, holds a rare trove of Persian books
of stones, including manuscripts of the *Javahirnama* of Muhammad ibn
Mansur, a widely circulating mid-fifteenth-century text that the imperial workshops of the Safavids, Qajars, Mughals, and Ottomans commonly reproduced from early modern times through the nineteenth century. Still others may be found in the manuscript collection of Edward
G. Browne (1862–1926), a professor of Arabic and Persian at Pem-
broke College in Cambridge University. Browne acquired these geological manuscripts, remnants of the pursuits of nineteenth-century
British Orientalism, from Albert Houtum-Schindler, a German scholar
of Persian and former superintendent of the turquoise mines in Iran.
The books of stones in the Browne Collection at Cambridge are records
of the lost story of the Qajar dynasty's ill-fated attempts in the nineteenth century to reclaim and revive the old mines of Nishapur, which
had once exported the finest turquoise stones in the world. Among the
Browne manuscripts, which were once shelved in the libraries of nineteenth-century Persian princes, are copies of books of precious stones
commissioned by Hulagu Khan of the Mongol Ilkhanid dynasty, the
Timurid emperor Shah Rukh, the ruler of the White Sheep dynasty
Uzun Hasan, the Ottoman sultan Selim I, and the Mughal emperor
Humayun, spanning the breadth of Islamic Eurasia.

The *javahirnama* genre provides a unique record of the trade in earth
substances and the exchange of the knowledge of stones among the
early modern tributary empires of Islamic Eurasia. These leather-bound
volumes in Persian nastaliq script enumerate and describe the precious
stones and mineral substances found on earth. Following an introduc-

tory preamble on the elements, the origins of matter in the universe, and the formation of stones on earth, *javahirnama* manuscripts consider each type of precious stone in its own chapter, delineating its origins, properties, and values. Scholars working predominantly with company records have often said that Asian indigenous source materials yield little information on pressing social or economic questions. However, vernacular materials present different sorts of information. The *javahirnama* literature may not supply quantitative data on, for instance, how many bales of turquoise were shipped in given years—unlike the trade records on silks and spices of the colonial East India companies or of diaspora merchant networks—but as indigenous sources they recover the cultural meanings behind the mining of stones and their exchange as imperial objects across Islamic Eurasia. Taking measure of Houtum-Schindler's library of rare Persian works on mining and precious stones, Browne observed that he had never before seen a collection of manuscripts "so well chosen for a definite purpose of study."[26]

In catalogues of Persian manuscripts, books of precious stones are buried at the end and designated as *occult,* a term they use to refer to miscellaneous works on scientific and naturalist subjects—including books of horses (*farasnama*) and falcons (*baznama*), for instance.[27] Such materials have been dismissed by an entrenched Orientalist tradition that emphasizes Islamic "philosophy," deems them unworthy of the canon of the classical Muslim *falsuf* (philosophers), and still remains hesitant, with certain exceptions, to engage with the framework of Islamic science. Meanwhile, historians of trade and commerce in the Near East and South Asia have yet to incorporate such works of indigenous natural history into their research because they seem arcane and tangential to useful social and economic data. The following pages attempt to read these sources with openness to the various possibilities and routes they offer for understanding the Eurasian turquoise trade, as opposed to expecting them to conform to standard social, economic, and political categories and preexisting frameworks.

Persian books of jewels offer a different, vernacular view of economic exchange in Iran, India, and Central Asia from the early modern period through the nineteenth century. They provide a window onto the cultural meanings and significance attached to stones such as the turquoise in the tributary economies of Islamic empires. These naturalist accounts and compilations cover the spectrum of the known precious stones of the world from their unearthing at the mines to the passage and exchange of stones and their material culture along networks of trade, tribute, and plunder.

The Persian source materials are complemented in this book by an array of early modern European printed geological and mineralogical texts from the Hoover Collection of Mining and Metallurgy in Honnold Library on the campus of Pomona College and at the Huntington Library in Pasadena, California. This voluminous literature, consisting of lapidaries, jewelers' manuals, and geologies on the sources, characteristics, value, and uses of mineral substances from around the world, was printed between the sixteenth and nineteenth centuries amid an unprecedented global traffic in precious stones, an outgrowth of imperial prospecting for natural resources—plants, spices, and stones—of commercial and medicinal value. Like the Perso-Islamic genre of javahirnama, European mineralogical literature appeared in the context of cross-cultural encounters and economic exchanges. Taken together, these natural histories of stones in Persian and European languages convey pieces of the lost history of the Eurasian turquoise trade.

TURQUOISE: THE SKY BLUE STONE

Turquoise, a phosphate of aluminum and copper, forms over deep geological time as crystals of feldspar and aluminum in the seams between subterranean folds of igneous rocks come into contact with veins of copper. The stone was linked to copper working and was first extracted from the earth in the Sinai Peninsula, at the deposits of Wadi Maghara and Sarabit al-Khadim, which were depleted around the twelfth century B.C.E. The turquoise trade subsequently moved eastward, coming to center on the caravan routes of Iran, Central Asia, and India and the city of Nishapur in eastern Iran, the home of the famed Abu Ishaqi (or ʿAbd al-Razzaqi) mine and its brilliant sky-blue stones. While the antiquity of the Nishapur mines cannot be determined, they were in operation by the tenth century C.E.[28] Local miners extracted their turquoise from time to time but not in any extensive way. Then, in the age of the post-Mongol Eurasian world empires, everything about the trade and culture of turquoise changed.

Under the Timurid (1370–1510) and Safavid (1501–1722) dynasties, the turquoise mines of Nishapur reached their peak of production, and the stone became an object of imperial conquest and the Eurasian caravan trade. It is most likely because of this imperial connection—and the long-held belief that a rider carrying a piece of turquoise would never fall from his horse or see defeat—that the Persian word for turquoise (*firuza*) shares a root with the word for victory (*piruzi*). In the Nishapur mines, local villagers unearthed turquoise still in its rock matrix, while above ground, middlemen purchased the stones from the miners and traded them to the gem merchants in the nearby shrine city of Mashhad. These merchants sorted, cut, and polished the turquoise before it traveled along overland caravan routes across Central Asia, South Asia, and the Near East. From the seven turquoise mines of Nishapur, sky-blue stones were traded, gifted, and looted—objects of exchange among the Timurids, Safavids, Muhgals, and Ottomans. Turquoise, stone and color, became immersed in the tributary economies of early modern Islamic empires in Iran, India, and Central Asia, where the exchange of objects underlined the mediation of imperial power and sovereignty.

Thus, the first Safavid monarch, Shah Isma'il I, famously sent turquoise stones as gifts to Isma'il Adil Shah of Goa in 1513 and to the Portuguese admiral Alphonse de Albuquerque the following year. In 1544, following his period of exile in the court of the Safavid shah Tahmasp, the Mughal emperor Humayun visited the turquoise mines of Nishapur on his way back to India, taking sky-blue stones with him as he set out to reconquer his empire from the Suri Afghans. In the early sixteenth century, the exchange of turquoise as an imperial gift became entangled in the Safavid and Mughal contention over the Afghan city of Qandahar, which lay between the two empires. The stone also spread through wars and plunder. Following the Battle of Chaldiran, in 1514, when the gunpowder-equipped Ottoman army crushed the mounted Qizilbash archers of the Safavids, the victors looted innumerable objects and artifacts, including turquoise-bejeweled objects, from Shah Isma'il's collection in Tabriz and took them back to the Topkapi Sarayi in Istanbul.

In addition to the physical object, the turquoise trade circulated the culture of the stone. Objects carry meanings, and as turquoise traveled, it left its meanings behind. The material culture of the stone was etched in Persian natural histories and books of precious stones, written in the context of imperial encounters with environments and the earth's natural resources as empires strove to order and convert the world of minerals, a subterranean terra incognita. These natural histories were exchanged in the course of interactions among the early modern Islamic empires of Eurasia, which shared Central Asian and Persianate cultural connections. The most widespread meanings that turquoise took on were its function as a stone of imperial victory and power and as a celestial shade of blue.

These Central Asian and Persianate cultural currents may be traced through an exploration of urban space and imperial architecture. From the fifteenth century, when turquoise emerged as a valued global object and commodity, blue was the predominant and most visible color of metropolitan style and architectural structures in the new imperial capitals and oasis cities located along the trade routes of the eastern Islamic world—Timurid Samarqand and Herat, Tabriz of the White and the Black Sheep dynasties, Safavid Isfahan, Thatta and Hyderabad in the Mughal province of Sindh. Tile makers and ceramicists replicated the seven celestial shades—turquoise blue, night blue, black, green, red, ocher, and white—of the *haft rang*, rooted in Persian mystical romances, to color the urban architecture of Eurasia. In the workshops of Eurasian oasis cities, artists and craftsmen mixed roasted copper, lead, and tin to produce the shade of firuza. From cobalt and roasted copper came the darker azure of *lajvard*. Set on the domes, minarets, portals, and surfaces of mosques and other monuments, tiles fired with the turquoise blue of Islam adorned imperial cities and urban spaces across early modern Eurasia.

Through channels of early modern travel and trade, particularly overland through the Ottoman Empire, turquoise reached Europe, an exotic object and global commodity from the distant Orient. In the cities of "Farangistan," turquoises with favorable size, shape, and—most important—color were classified as ring stones and sold by the piece, while smaller stones were merchandised by weight. In early modern Europe, where jewelers put it in rings, turquoise was classified as a mineral substance the color of the sky, with which shade it became synonymous. It was one of the mineral substances mined in early modern Iran and Afghanistan that was valued for its chromatic effects, and it gave substance to the perception of blue as one of nature's colors. Another was lapis lazuli, a darker blue stone from Badakhshan Province in the mountains of northern Afghanistan, the source of the esteemed pigment *ultramarino,* prized by the painters of the Renaissance, and the root of the word *azure*. The trade of turquoise and lapis lazuli, along with that of cobalt, a mineral found in central Iran used to color Chinese porcelain, carried the material culture of blue from Asia to Europe.

But turquoise took on a different purchase and cultural value among European trading companies, commercial empires, and consumers, to whom it was one of many exotic objects circulating around the globe. Unlike in the early modern Near East, Central Asia, and South Asia, where it was gifted, traded, and read as a particular symbol of conquest and empire, in European markets, where the more malleable metals of gold and silver were the standards, the stone was demystified and lost its symbolic value. Its monetary worth also diminished as it moved from natural object to global commodity, from a precious to a semiprecious stone. Turquoise's Eurasian trade and meanings faded amid the divergence between the tributary economies of the post-Timurid Islamic empires and the commercial economy of the early modern period. Different economic values and demands for different gems and mineral substances emerged as European empires gained access to colonies and the resources of the Indies and the New World.[29] The turbulence and dynastic instability that followed the fall of the Safavid Empire in 1722 also dealt a heavy blow to the Eurasian turquoise trade. Over time, the resulting deterioration of mining standards and the ruin of many of the most valuable mines, over which the empire had held a monopoly, effectively brought about the demise of the Eurasian turquoise trade and the stone's culture as a symbol of victory and imperial power.

In the nineteenth century, an age of gold rushes, empires and prospectors rediscovered and attempted to restore ruined turquoise mines, seeking to reclaim desert nature and profit from the stone's global commerce. In the 1840s, following on the heels of late eighteenth- and early nineteenth-century European exploration of ancient Egypt's ruins, monuments, and environments, the British cavalry officer Charles Macdonald set out to reopen the Sinai turquoise mines. He moved out to the land of turquoise with his family, built a house of stone, enlisted the local Bedouin, whom he equipped with gunpowder for mining, and strove for years to revive the buried mines of the ancient Egyptians. In 1882, on his second tour of the eastern frontier province of Khurasan, Nasir al-Din Shah of Qajar Iran schemed to revive the turquoise trade that had once thrived under the Safavid dynasty by reestablishing state regulation over the mines of Nishapur. Encouraged by reports in Persian gazetteers of the gold rush in the American West and by histories of the New World in circulation at the time, the shah and his ministers strove to reclaim the turquoise mines and to corner the global commerce in the stones.

Meanwhile, on the other side of the world, accessible sources of turquoise in the Americas—the lost mines that had once supplied the regalia of the Aztec Empire—reopened, including the famed Cerrillos Mines of New Mexico, ending the Persian monopoly and transforming the patterns and significance of the turquoise trade. The restoration and growth of the turquoise industry in the American Southwest in the 1890s paralleled the waning of Eurasian traffic in the stone. Turquoise was no longer an exotic or rarity attainable only from distant Asian mines. The turquoise trade that had once spanned Islamic Eurasia and shaped imperial encounters while giving the world the color sky blue had lost its meaning.

CHAPTER I

The Colored Earth

FLOWERS OF COPPER

From the depths of the earth comes a piece of the sky. Turquoise is a mineral substance formed in the seams of rocks below the surface of the earth. A hydrous phosphate of copper and aluminum, it is born in igneous rocks, as magma, fiery liquid deep within the earth, surges toward the surface, pools, and solidifies. In a geological process that lasts thousands of years, nature weathers, buries, and erodes these rocks, bringing their copper, aluminum, phosphorus, oxygen, hydrogen, and water together to create the chemistry of turquoise, $CuAl_6(PO_4)_4(OH)_8 \cdot 4H_2O$ (see table 1).[1] An opaque stone whose pale blue results from the presence of copper, turquoise has surfaced in desert environments from Eurasia to Mesoamerica, been valued as a celestial stone, and been traded across the world.

A nineteenth-century Persian natural history compares the formation of turquoise to the ripening of fruit: "It is said turquoise is like a cherry—the more it ripens, the better. But all the cherry needs to ripen is the sun of one season, whereas for turquoise it takes one thousand years [*yak hazar sal mudat lazim hast*]."[2] To reach the state at which it can be cut from its rock matrix and traded as a precious stone and ornament, it must undergo deep geological alterations in the layers of the earth. If turquoise is mined before it has aged, its color fades.

Igneous rocks rich in aluminum and copper minerals give birth to turquoise. Over centuries, deep-seated weathering and exposure to the

TABLE I   TRUE TURQUOISE

Chemical Composition	
Phosphorous oxide	32.8 percent
Alumina	40.2 percent
Water	19.2 percent
Copper oxide	5.3 percent
Iron and manganese oxide	2.5 percent
Total	100.0 percent
Hardness	6
Specific Gravity	2.75
Form	Amorphous
Luminance	Opaque

elements alter their form. As rainwater seeps from the surface through these rocks, it breaks down and converts their minerals into new chemical substances.[3] Waters containing phosphoric acid that flow through aluminous rocks decompose feldspar crystals into kaolin, a mineral substance that is associated with turquoise, freeing necessary aluminum silicates.[4] All of this slow, nearly motionless change occurs in the zone of oxidation, in the first few hundred feet below the crust of the earth.

As these minerals solidify into rocks, they enclose the now crystallized turquoise, a naturally occurring mineral substance with a definite chemical composition.[5] Contact with copper ores and the passing of time give turquoise, an opaque stone with a soft surface and a high index of light refraction, its defining physical property—a sky-blue color. While turquoise can have many shades, including a common pale green caused by traces of iron, the finest stones, deemed gems—those aged for thousands of years in the oldest rocks—are sky blue. But the color is inconstant and unstable and can fade to green. It is in the nature of turquoise to change. When exposed to the elements in the open air, turquoise weathers, decomposing into a dusty white powder.[6]

Turquoise was first found after wind and rain had eroded the surface rocks above it. Thus freed, turquoise stones in the open alluvial gravels at the foot of mountains and blown across deserts by high winds caught the eyes of passersby and came to be regarded as valuable natural objects. Because turquoise could be found only a few hundred feet below ground and was easily extracted, miners descended into caves and dug for it with rough tools, including shovels, pickaxes, and hammers, separating the blue stones from the veins of rocks. Windlasses raised the mineral, still encased in its matrix, from the mines. Removed from these broken pieces of rock, turquoise stones were cut and polished into gems and traded across the Eurasian expanse.

DESERT NATURE

The chemistry of turquoise occurs in desert environments from Mesoamerica to Eurasia. For centuries, the hub of its trade was Asia, where turquoise was unearthed in a mineral-bearing stratum extending from the Sinai in Egypt through Iran to Tibet, with the city of Nishapur in eastern Iran being the historic heartland for the finest stones, those the color of the sky. In Mesoamerica, turquoise (called *chalchihuitl*), the best of it mined from the Cerrillos district of New Mexico, became an object of tribute and sacred regalia in the Aztec Empire in the postclassical period (c. 900–1521 C.E.). Spanish explorers described a blue-green stone that the indigenous populations treasured, and in 1519, Montezuma II famously offered gifts of turquoise to the conquistador Hernán Cortés. The stone was first unearthed, however, thousands of years earlier on the other side of the world, in the predynastic Egyptian Sinai.[7]

The turquoise mines of the Sinai materialized in sandstone ridges connected to the working of nearby copper mines and were known to the Egyptians before 5500 B.C.E.[8] According to one Egyptologist, "The earliest signs in Egypt of intercourse with Sinai are the beads of turquoise" found among the remnants in early dynastic graves.[9] The Monitu Bedouin clans of the Sinai likely first unearthed turquoise, known in ancient Egypt as *mafkat,* but not to any great extent. The stone became known as it was carried along the caravan routes that extended from Egypt across Asia and Africa.[10] The pharaohs sent expeditions into the Sinai to secure mineral deposits necessary for their monumental building projects in the Nile Delta.[11] Seekers of turquoise entered a distant and unforgiving environment, where they were forced to pay tribute to the local Bedouin populations and to protect their mining works from the raids of the fiercely independent Monitu, referred to in inscriptions as "Lords of the Sands."[12]

Numerous inscriptions, stelae, and graffiti record the details of mining expeditions and attest to dynastic exploits in mineral extraction.[13] The turquoise mines in Wadi Maghara, or the Valley of Caves (or Grottoes), were the first to be exploited, as inscriptions found there record.

These range from those of Sneferu (r. c. 2613–2589 B.C.E.), of the Fourth Dynasty, who recorded his conquest of the country and his discovery of the mines, to those detailing an expedition sent by Thutmose III, of the Eighteenth Dynasty. Captives worked the mines for more than two thousand years, the marks of their metal tools chiseled into the steep sandstone walls and cavernous rocks of Wadi Maghara.[14]

The age of the legendary Sneferu was especially connected with the Sinai, as indicated by a bas-relief on the northwestern slopes of Wadi Maghara depicting a Bedouin chief in a subservient pose before the Egyptian pharaoh. Sneferu's haul in turquoise was regarded as exceptional. An Egyptian tale about a lost jewel of "new turquoise" recovered by magic suggests the stone's value in Egypt during Sneferu's reign, when it was an object in patterns of wider economic and cultural consumption, ornamenting bodies and stories.[15] Following the reign of Pepy II, circa 2185 B.C.E., the mining colonies in Wadi Maghara were "abandoned" and "lingered in comparative idleness, their veins of turquoise seemingly exhausted."[16]

But the demand for the stone remained great, and explorers soon discovered untouched turquoise deposits about ten miles away, at what is now Sarabit al-Khadim, or Heights of the Servant, named in reference to a statue of a slave carved in black stone, which the French are said to have carried away during their occupation of Egypt. These mines were also likely opened in the time of Sneferu, as inscriptions record his presence there.[17] Starting in the reign of Amenemhat II (r. c. 1929–1895 B.C.E.), successive expeditions worked and exploited these new veins.

The workers dedicated the mines to Hathor, the goddess of the turquoise land, the goddess of turquoise, and built temples in her honor.[18] She was represented as a moon-faced woman carved on a stone column. Near the mines at Sarabit was a shrine (*hanafiya*) of the Turquoise Goddess, where prayers, vows, and offerings could be made. The demand for new veins of turquoise led Queen Hatshepsut (c. 1508–1458 B.C.E.) to order the reopening of the Wadi Maghara mines, which had not been worked for nearly four hundred years, and she sent an agent "with orders to inspect the valleys, examine the veins, and restore there the temple of the goddess Hathor."[19] Akhenaten in the Amarna Period (1353–1336 B.C.E.) and Rameses III of the Twentieth Dynasty (c. 1186–1069 B.C.E.) sent several expeditions to the turquoise mines in the Sinai.[20] The last pharaoh to leave an inscription at Sarabit al-Khadim was Ramses VI (r. c. 1145–1137 B.C.E.), and the rocks have no further traces of mining after that.[21] The turquoise mines were exhausted and abandoned, as was the temple of the turquoise goddess Hathor after the Twentieth Dynasty.

By the medieval period, the geography of turquoise and its trade had shifted to West, Central, and South Asia—to the Persianate world, where turquoise was known as *firuza*. The stone garnered such high esteem there that it became the name of victorious dynasts and of cities and mountains where it was not even to be found. The legendary lost kingdom of Firuzkuh, or Turquoise Mountain, thought to have been near the blue-tiled Minaret of Jam in the Hindu Kush of northwestern Afghanistan, was associated with the stone before the Mongols leveled it in the thirteenth century. Although there are no major turquoise deposits in its vicinity, northern Afghanistan being better known for the rich deposits of balas rubies and lapis lazuli from the mines of Badakhshan, the sultans of the Ghurid Empire (c. 1100–1215) founded and envisioned Firuzkuh as a city of turquoise, the center point of a network of fortified settlements. The Ghurids built their monumental imperial capital as a display of power and as a treasury to store the spoils of victorious wars in India and Iran.[22] The source of inspiration for the now-lost city of Turquoise Mountain could be found in the blue stones mined to the west, outside the city of Nishapur.

THE TURQUOISE CITY

Nishapur is in a vast plain at the foot of the Binalud Mountains in the eastern Iranian province of Khurasan. For more than a thousand years, these mountains have been the most important source of turquoise in the world. Along with Herat, Balkh, and Marv, the medieval city of Nishapur was once among the most important urban centers in western and central Asia.[23] It was a thriving oasis and entrepôt along the Silk Roads and a cultural hub that was the birthplace of such Perso-Islamic figures as the mathematician and philosopher Omar Khayyam and the Sufi poet Farid al-Din 'Attar. In the twelfth and thirteenth centuries, repeated earthquakes, combined with the Mongols' sudden and violent conquest of the city in 1221, reduced Nishapur to a modest provincial town. Devastating earthquakes in 1209 and 1270 destroyed the first two sites of the city, which had to be moved and rebuilt.[24] Despite this decline from its former stature, just a few miles north of Nishapur were still the most valuable turquoise mines on earth and the hub of the world turquoise trade.

By the thirteenth century, turquoise was mined in six valleys about thirty-five miles north of Nishapur in the villages of Ma'dan. The most famous of these was ʿAbd al-Razzaqi, containing an old and extensive mine of the same name (also known as Abu Ishaqi, or Isaac's Mine), so esteemed for the quality of its stones that the Persian poet Hafiz hinted at it in a verse on the mutability of the earthly world: "In truth the turquoise ring of Abu Ishaq / Flashed finely but then faded away."[25] The stones were taken from the mines to the nearby cities of Nishapur and Mashhad, where craftsmen cut them, pasted them onto strips of wood, and polished them on a grinding wheel. Shined with pieces of leather, the stones were set in rings, inlaid in metal objects, and arranged as tesserae, then brought to markets across Asia and the Near East.[26]

Classical Arab and Persian geographers identified the mountains of Nishapur as the source of this turquoise. In the tenth-century *Ahsan al-Taqasim fi Maʿrifat al-Aqalim* (The best of the divisions of the knowledge of regions), the Arab geographer al-Muqaddasi (c. 945–91) accounts turquoise, which he calls *fayruziyya,* the main export of Nishapur, noting that the city's stones were used to ornament the prayer niches (*mihrab*) of mosques as far away as Damascus.[27] Abu Mansur al-Thaʿalibi (961–1038), a literary figure who hailed from Nishapur, praised its exceptional turquoises and edible earth.[28] Writing in the mid-fourteenth century, Hamdallah Mustawfi (1281–1349) left an account of precious stones in his world geography *Nuzhat al-Qulub.* Of turquoise, he wrote, "There are many mines of this stone, but the best mine is that of Nishapur, by reason of the good quality of the stones and the little labor in getting them. In the mountains of Nishapur there are pits dug where the turquoises are found, and thence come the best stones."[29] Still, Mustawfi reported that the mines had to be maintained or else could fall into disuse, thus affecting the trade and circulation of the stones: "These Nishapur turquoises were famous; but of late years, scorpions have come to be found in these pits, and in fear of them people have ceased to work the mines."[30]

FINDING TURQUOISE

The most comprehensive sources on turquoise and its early trade are medieval Arabic and Persian books of precious stones. This genre of books, which the imperial courts of Islamic Eurasia regularly sponsored, classifies the mineral substances of the earth and their properties, including their value as jewels and materia medica. One of the earliest surviving examples is *Kitab al-Jamahir fi Maʿrifat al-Jawahir* (Compendium on the knowledge of precious stones), an eleventh-century Arabic work of cosmography and mineralogy attributed to Abu Rayhan al-Biruni (973–1050), a Central Asian Muslim scholar and traveler in the service of the Ghaznavid Empire, in what is now Afghanistan. In this treatise on rare objects found in the mines of the earth and the treasuries of kings, al-Biruni promises "the description, categorization, and assessment of the precious objects and jewels that lay buried as treasures."[31] He refers to and frequently cites from an earlier book on jewels and "treasures interred in the earth," by the Kufan philosopher and polymath Abu Yusuf Ya'qub bin Ishaq al-Kindi (c. 801–66).[32] Al-Kindi's text is one of a number of lost early Arabic scientific books on precious stones, with others including the anonymous and undated *Durar al-Kamina* (Hidden pearls) and *Kinz al-Tujjar fi Ma'rifat al-Ahjar* (The treasury of merchants and the knowledge of stones) and the many works of the eighth-century scholar Jabir al-Kharaqi, such as *Kitab al-Ahjar* (Book of stones) and *Rasa'il fil-Hajjar* (Treatise on rocks and stones). Another lost text, *Rasa'il Ba'ad al-Hukama wa-l-'Ulama al-Qudama' fil-Jawahir wa-l-Khawass* (Treatises written by ancient philosophers and scholars on precious stones and their properties), is attributed to the Bukharan scientist and philosopher Abu 'Ali al-Husayn ibn 'Abdallah ibn Sina (c. 980–1037), known in the West as Avicenna.[33]

In the thirteenth century, Ahmad ibn Yusuf al-Tifaschi (1184–1253), a jeweler and scholar from Cairo, thoroughly described the formation, geography, properties, and values of twenty-five types of precious stone. His *Azhar al-Afkar fi Jawahir al-Ahjar* (Best thoughts on the best of stones), written in the days of the Ayyubid dynasty (1171–1250) in Egypt, on the eve of the rise of the Mamluk dynasty (1250–1517), promises "strange information . . . of great benefit" on "a variety of precious stones that no great king or important nobleman can do without in view of their unusual benefits and great properties."[34]

Works in the Persian genre of *javahirnama* (books of precious stones) took the measure of the minerals and metals of the earth even further over the course of the medieval and early modern periods. The *Javahir-nama-yi Nizami* (Ordered book of precious stones) by Muhammad bin Abi Barakat Javahiri Nishapuri, a jewel maker and merchant of stones from Nishapur, is a compendium of knowledge on metals, minerals, and precious stones written circa 1196 and perhaps the earliest extant work on minerals and precious stones in the Persian language. It is the presumed source of the later, Mongol-era Persian text *Tansuqnama-yi Ilkhani* (The Ilkhanid book of rarities, gifts, and tributes), written circa 1265 by the Muslim astronomer and savant Nasir al-Din Tusi, and

Abu'l Qasim 'Abdallah Kashani's fourteenth-century '*Ara'is al-Javahir va Nafa'is al-Ata'ib* (Statements on jewels and gifts of rarities).[35]

These Perso-Islamic works of natural history hold the four principle elements ('*anasir*)—earth (*khak*), water (*ab*), air (*hava*), and fire (*atash*)—to be the source of everything in the world.[36] According to Islamic scientific traditions related in books of precious stones, "the rays of the sun [*aftab*] and stars [*kavakib*] and the heat from celestial bodies combined with the coldness of earth to form stones. Mines are born in this way."[37] The four principle elements, also known as *chahar unsurb*, described the earth's relation to the universe: earths were the base, water was on their surface, air was on the surface of water, and fire loomed above. Unusual changes in environmental conditions—such as exposure to heat (*hararat*) and cold (*barudat*) and smoke and vapors (*bukhar*)—were thought to form mines, creating mineral deposits in the mountains.[38] These precious mineral substances were moved across high deserts by winds and washed up in the beds of rivers, where they were found.

Islamic books of stones from the medieval period described all the known gems of the world, identifying their origins, qualities, and trade. They praised turquoise (*firuza* or *fayruz*) for the lustrous color that was the essence of its value and that brought it into high demand across the lands of Islam. One of the earliest descriptions of its attributes and value appeared in al-Biruni's *Kitab al-Jamahir*: "A bluish stone, harder than lapis lazuli . . . mined from the mountain of San in Khan Ruyand [Nishapur]. If rubbed on a rock or stone after dilution with water, it will readily accept moisture. It is then oiled and filed so as to be made soft. The more humid it is, the better it would be. In the course of time it gains sharpness and color."[39] The "best kind," al-Biruni wrote, was mined from the Azhari and Bu Sahaqi mines of Nishapur and fetched a price of ten dinars for one dirham of stone.[40]

Subsequent Islamic books of precious stones written in Persian went much deeper in their accounts of the turquoise of Nishapur and its trade. The author of *Javahirnama-yi Nizami* hailed from Nishapur and gathered his information directly from the turquoise mines while also drawing extensively on his wide travels, including journeys to the ports of the western Indian Ocean, to detail the mining and trade of precious stones and minerals. This jeweler's compendium provides an account of the economy and culture of the Eurasian turquoise trade. According to Nishapuri, there were only four known deposits of turquoise on earth, in the vicinity of Nishapur and in the Central Asian lands of Khvarazm, Transoxiana, and Turkistan. Stonecutters and jewelers could discern from which mine a stone originated when they saw it.[41] The premium stones came from seven famed mines outside Nishapur, which were identified by medieval Persian mineralogical texts. They noted that the bluest, most brilliantly colored (*rangin*) and coveted stones were unearthed at the Abu Ishaqi mine outside the city. The other Nishapur mines known to contain fine blue turquoise, in the order in which the books treat them, were Azhari, Sulaymani, Zarhuni (whose turquoise was streaked with gold), 'Abd al-Majidi, Andalibi, and Asuman Gun (or Khaki).[42] In *Tansuqnama-yi Ilkhani*, Tusi links the color of the stone to the sky and air (*hava*), noting that a dustless, clear blue firmament gives the stone more radiance (*lun*).[43] Adding to the mystery of turquoise was the fact that its color could change, even fade to green, lowering the stone's value. Medieval authors, such as al-Biruni, attributed this instability to the effects of atmosphere and climate—cloudy skies, winds, and sun—on the stone.[44] Because of this unstable nature, some people would have nothing to do with turquoise. When a stone lost its color, it lost its worth as well, being known in the language of jewelers as a dead stone (*sang-i murda*), although it could be brought back to life through contact with water and sun.[45]

The blue of turquoise transcended its physical properties, and folkloric beliefs attached to the stone and its celestial hues. A reflection of the ethereal sky found in the dusty earth, turquoise held a marvelous appeal. In bazaars and markets across Asia, it was priced by color and weight. The *Javahirnama-yi Nizami* notes that in Nishapur circa 1200, one dirham of turquoise was worth sixty dinars.[46] The most valuable turquoise of the age, according to Nishapuri, was a stone from the Abu Ishaqi mine that weighed less than a dirham but was appraised at one hundred dinars because of its radiant luster and color.[47] It belonged to Sultan Sanjar bin Malikshah of the Saljuq dynasty, on the hand of whose wife, Khatun Ajjal, the author of *Javarinama-yi Nizami* saw the incomparable stone worn.[48] Writing in the following century, Tusi also left a record of the market value of turquoise: half a mithqal of Abu Ishaqi and Azhari stones was priced at seven to ten dinars, one mithqal at twenty to thirty dinars, two mithqal at fifty to seventy dinars, and three mithqal at one hundred to one hundred and fifty dinars. Lesser-quality turquoise could be procured at little cost.[49]

Traded along the Silk Roads, turquoise stones passed eastward across Asia, reaching "the people of Chin, Machin, Tughmakh, and Tangut" and becoming talismans and ornaments to be placed on Buddhist idols.[50] The trade also passed westward, reaching the Mediterranean world, where, according to al-Tifaschi's *Azhar al-Afkar,* the Berber princes of the Maghrib and their followers coveted the stone as adornment for swords and rings and paid ten Moroccan dinars for a single specimen from Nishapur.[51]

THE COLOR OF THE SKY

Some thousands of years ago, turquoise was born underneath the earth's crust in seams of rocks rich in aluminum and copper that weathered and aged over time until the mineral substance crystallized as an opaque sky-blue phosphate. Through erosion, turquoise separated from its outer rock and became exposed as a precious stone. During the Islamic period, turquoise spread from mines in the high desert of eastern Iran across Asia, traded as ornament, talisman, and jewel.

By the fifteenth century, the turquoise trade—a transmission of stone and color—had entered into currents and networks of Eurasian cross-cultural exchange unprecedented in distance or volume, and its traffic acquired new purchase across the Near East, Central Asia, and South Asia.[52] Islamic courts in early modern Eurasia raised turquoise into an object of tribute and interimperial rivalry and exchange.[53] The stone evolved into a substance in the display of conquest and power among Eurasian empires, reflected in sky blue imperial cities and their monuments. Reaching early modern Europe as one of the exotic, colorful objects and luxuries to be collected from the East, it sparked the cultural construction of the color blue. So unique was its hue in nature that *turquoise* became the classification of its signature shade.

CHAPTER 2

Turquoise, Trade, and Empire in Early Modern Eurasia

It was the stone of victory: *firuza*. The stone of the legendary lost turquoise thrones of Firdawsi's *Shahnama* (Book of kings), emulated in the blue-tiled *maydans* (esplanades) of oasis cities. To the Indo-Persian Islamic empires of early modern Eurasia, it was a sacred mineral substance and an object of interimperial exchange. From its remote beginnings, buried in mines outside the city of Nishapur in eastern Iran, turquoise and its material culture traveled across India, Central Asia, and the Near East and entered into the tributary economies of the post-Timurid empires of Islamic Eurasia. Turquoise became ingrained in the customs of the mediation and projection of imperial sovereignty and sway that prevailed in the court cultures of Turko-Mongol dynasties.

The history of Eurasian commodities and their worldwide circulation has been examined mostly through the prism of silk and its commerce. Thomas Allsen's *Commodity and Exchange in the Mongol Empire: A Cultural History of Islamic Textiles* explores how the Mongols built a transcontinental empire that expanded commercial trade along the Silk Road from East Asia across Central Asia and the Near East.[1] Resettling Muslim textile workers and artisans in China, where they made robes of golden silk brocade called *nasij* for the imperial court, the Mongol world empires, infamous for their destructiveness, strove to facilitate and open the channels of global commerce in silk and Islamic textiles. Turning to a later period, in *The Politics of Trade in Safavid Iran: Silk for Silver, 1600–1730*, Rudolph Matthee argues that the trade of silk—by all accounts the most commercially valuable commodity that the empire exported and "the mainstay" of its trade with Europe—spanned the whole of Iran's economic activity.[2] With the introduction and circulation in Europe of large quantities of silver from the Americas in the sixteenth century, the continent's expanding economies and trading companies sought and imported a diverse array of Asian commodities, such as silk, which Asia sold for the precious metals it lacked. Through an analysis of the expansion and decline of the silk trade under the Safavid Empire, Matthee details the commercial history of early modern Iran and its place in the world economy. Still, there was more to the economy and trade of Iran than silk, and recent studies have charted the production, circulation, and consumption of a wider array of Asian commodities while also highlighting different practices of exchange rooted in the tributary economies of Islamic empires.[3]

Building on these strands of literature, this chapter examines the history of turquoise in interimperial exchanges among early modern Eurasian tributary empires.[4] It parts ways with the long-standing emphasis on commercial economy and global capitalism to explore an inter-Asian vernacular economy and traffic that Europe and its trading companies could not absorb: the trade in turquoise stones. In the tributary networks of early modern Eurasian empires founded by Turkic pastoralists who created a synthesis of the steppe and the sown, linking the Central Asian and Persianate worlds, turquoise emerged as the stone of victory, the stone of conquest, and a primary element and theme of imperial regalia.

TURQUOISE EMPIRES

The Timurids (1370–1526) were a dynasty of Central Asian pastoral origins that made the transition to the culture of the Persianate world in the aftermath of the conquests of Timur Lang, known in the West as Tamerlane.[5] Expanding outward from their turquoise-tiled capital cities in Samarqand and Herat, they were the predecessors of the Safavid and Mughal Empires on the Iranian plateau and the Indian subcontinent and rivals of the Turkmen and Ottoman Empires in Anatolia.

Nishapur, the fabled turquoise city in the arid high plains of northeastern Iran, suffered much from adverse natural events in this period, when it was part of the Timurid Empire. After a series of chronic tremors that started in the twelfth century, in 1405, the first year of the reign of the Timurid ruler Shah Rukh (r. 1405–44), a devastating earthquake leveled the city. The contemporary historian and geographer Hafiz-i Abru (d. 1430) chronicled its effects: "A great earthquake [*zilzila-yi 'azim*] struck. The force of the earthquake was such that one imagined that no rocks were left on the mountains nor a clump of earth in the valleys, to the point that in open spaces the air was altered. It appeared that the compounds of the earth were turned as a result of the shaking of the planet. For several nights the earth shook and people took cover in a state of tumult. Apart from those in the surrounding desert, most of the other inhabitants of the city were buried in the ruins."[6]

Although it never fully recovered nor returned to its previous grand position, Nishapur was rebuilt and again thrived, though on a more provincial scale. Hafiz-i Abru praised the city's environment, its fresh air, the rivers and canals that supplied its water, and the earths that made up its mineral resources. The Binalud Mountains on its outskirts held vast deposits of the world's bluest turquoise: "In the mountains of Nishapur, there are mines of turquoise, and the good turquoise comes from there."[7] Despite the troubles it had undergone, the city remained the heartland of turquoise and the axis of the gem's global trade, yielding the most-sought-after turquoise in the world, known for its radiant sky-blue color.[8]

To meet this demand, workers toiled in the six turquoise valleys of the Binalud Mountains, mining turquoise according to the age-old methods. Using rough tools, they separated the turquoise from subterranean walls of rock and pulled stones, still in the rock matrix, from the mines by means of a windlass. Others searched for the mineral in washed-up disintegrated stones at the foot of mountains. Outside the mines, middlemen and merchants sorted through and purchased turquoises, which they transported to the nearby city of Mashhad. There, lapidaries (*hakkakan*) placed the stones on pieces of wood before cutting and polishing them on rough sandstone rocks (by the early nineteenth century, a sanding wheel turned by a bow and chord cut and polished turquoises). After this process, the stones were further polished with leather and turquoise dust and cut into different shapes before being exported across Eurasia.

In the Timurid period, turquoise acquired a broader ambit and significance than it had ever had before, moving around the world in unprecedented distance, scale, and volume. The culture of the Timurid Empire transformed turquoise into an object of interimperial contact and exchange, a measure of power and sovereignty, across Central Asia, South Asia, and the Near East, where the cerulean color of the stone was epitomized as the shade of monumental cities.

The world of the early modern Eurasian turquoise trade and its meanings can be entered through the Perso-Islamic genre of *javahir-nama*, or books of precious stones, sponsored by the Timurid, Safavid, Mughal, and Ottoman Empires. Muslim scholars drew on a rich tradition of medieval Arabic and Persian natural histories of precious stones and minerals known as *ahjar al-karima,* or the "great stones," to write these works, which classified the mineral substances of the earth, including their value as jewels and materia medica. Books of precious stones composed in the fifteenth century and later thus followed in the footsteps of long-established scientific genres about the earth and its substances, but they had a far wider circulation and impact than medieval texts and were commonly reproduced in the workshops of the interconnected Islamic empires of Asia and the Near East.

Zayn al-Din Muhammad Jami's *Mukhtasar dar bayan-i shinakhtan-i javahir* (Epitome on the recognition of gems), a fifteenth-century book of precious stones composed in Herat for Shah Rukh, details the Eurasian turquoise trade in the Timurid period.[9] The book is in twelve chapters, with each chapter (*bab*) on a particular type of stone, the sixth on turquoise. Under the Timurids, turquoise became emblematic in the fabric of imperial cultures across Central Asia, South Asia, and the Near East and reached European markets and consumers. Jami's book notes that turquoises, along with diamonds (*almas*), rubies (*yaqut*), pearls (*marvarid*), and other precious stones unearthed in Asia, were found in the markets of Farang (Europe) and gives their value in gold florins, the Italian coins struck in Florence and used widely in the channels of international commerce.[10] Jami, like his predecessors, confirmed that the choicest pieces of turquoise came from Nishapur, while newer, less-valuable stones, which often lost their color, were found in Khojand, Zanjan, and Kirman, other locales in Central Asia and Iran. Unique stones of high quality mined in Khojand sold on the market for the modest price of five florins. But the premium and true (*asli*) turquoise came from the Abu Ishaqi mine of Nishapur. An Abu Ishaqi stone of sky blue weighing roughly twenty carats (*qirat*) would cost four hundred florins, while smaller stones sold for up to fifty florins per carat.[11]

Jami deemed turquoise the stone with the most excellent properties (*dar khasiyat bihtarin ahjar ast*), suggesting the importance that Timurid culture attached to it.[12] It carried imperial connotations and was believed to bring courage and victory (*piruz*) over enemies in war. If worn by rulers, it turned their wrath into mercy toward their subjects. Sultans dearly esteemed and vied for the royal treasuries that were stocked with the Abu Ishaqi variety of the stone. The people (*jami' al-khala'iq*) regarded turquoise as auspicious to look upon, protecting wearers from harm, portending a long life, and preventing nightmares. Oculists used the stone to treat diseases of the eye.[13]

Turquoise also carried much esteem among the Timurids' rivals to the west, the Qara Quyunlu (Black Sheep) and the Aq Quyunlu (White Sheep).[14] These Turkmen dynasties, straddling the eastern Anatolian frontier in Azerbaijan and Diyarbakir, were likely among the conduits through which turquoise mined in Timurid Nishapur reached the markets of the Ottoman Empire and Europe. The Qara Quyunlu (1380–1468) and the Aq Quyunlu (1378–1508) rose to power in eastern Anatolia and northern Iraq and expanded into western Iran, becoming centered in the capital city of Tabriz—repeating the pattern of Turkic pastoral nomadic confederations moving from the tent to the throne. Under Abu'l Muzaffar Jahan Shah (r. 1438–67), the Qara Quyunlu reached their peak and Tabriz signaled the westward transmission of Timurid metropolitan culture.[15] Jahan Shah ordered the construction of the Blue Mosque of Tabriz, known as Firuza-yi Islam, the Turquoise of Islam, in line with the imperial architecture of Timurid Samarqand and Herat, distinguished by the radiant shades of glazed tiles fired in the "seven colors" of heaven, two of which were turquoise and night blue. In 1468, when the Aq Quyunlu clans defeated the Qara Quyunlu in battle, dethroning Jahan Shah, they and their commander Amir Hasan Bayg (r. 1453–78), better known as Uzun Hasan, assumed power in the capital city of Tabriz.[16]

BOOKS OF STONES

In order to identify the rare objects and gems collected in the treasury of Uzun Hasan, the Aq Quyunlu sponsored the composition of what became the fullest and most widely circulating early modern text on mineralogy and precious stones in Islamic literature. The White Sheep *Javahirnama* (Book of precious stones) was "written in the name of Khalil Bahadur Sultan, slave of Uzun Hasan"—denoting its composition by order of Uzun Hasan's son—by Muhammad ibn Mansur al-Shirazi (1425–98), a scholar from the learned Dashtaki family of the city of Shiraz in the southern Iranian province of Fars.[17] Commonly reproduced in the imperial workshops of the Mughals, the Safavids, and the Qajars between the fifteenth and nineteenth centuries, Ibn Mansur's *Javahirnama* offered a compendium of existing knowledge of precious stones in the Islamic tradition and was recognized as the most comprehensive work in its namesake genre. The text provides scientific and economic knowledge of the mineralogical objects encountered and exchanged by early modern Islamic empires. Its opening metaphorically suggests this combined interest in the science and the trade of stones by setting upon the exploration of "the bazaar of the universe . . . [under] the dinar of the sun and the dirham of stars."[18]

Early modern Persianate books of precious stones ostensibly attribute the existence of elements, minerals, and gems in the earth to the firmament and the divine. Ibn Mansur's *Javahirnama* begins with a description of the creation of the universe, the stars, and the earth and its wonders:

> Infinite praise be unto the Creator [*hakim*], who placed the moving and motionless stars into the bazaar of the universe. The one who set the dinar of the sun and the dirham of stars into motion. The one who made the universe long and rendered the heart of the sage full of knowledge. The hakim, who with the food of love of oxygen, mixed the matter of pure souls with water and earth, and from the blending of these brought human beings, the unique miracles of the world, into existence from nothingness, dispatching them to the house of dust. Earth was passed through the celestial universe, becoming visible, as days were lit with rays of the sun and the darkness of night with bright stars. With a stroke of the pen [*ba qalam*], he brought water into creation and painted colors on the earth. . . . The head of the caravan of the passengers of existence [*qafila salar-i musafiran-i vujud*], with the knowledge of alchemists, turned the copper of the earth to gold.[19]

Ibn Mansur's references to the connections between earth and sky belie a thoroughly naturalist discourse on the physical formation of rocks and stones on earth. It is nature and its geological processes that possess powers almost divine. "The creations of human crafts [*sana'at*] fall short of the creations of nature [*tabi'at*]," he writes, citing as support the tenth-century Muslim philosopher Ibn Sina's claim that "alchemists [*chimia garan*] do not have the power to control the essence of a thing but can only come to understand it and change its characteristics."[20] Ibn Mansur's speculations about the nature of rock formation were consistent with the geological knowledge of his times.

In a contemporary Arabic text, *Ma'din al-Nawadir fi Ma'rifat al-Jawahir* (A mine of anecdotes on the knowledge of jewels), by 'Ala' ibn al-Husayn Bayhaqi (d. 1509), the four principle elements ('*anasir*) of earth, water, air, and fire are the basis of an explanatory account of the formation of all matter, including stones: "We are acquainted with the movement of the sun in its oblique orbit, as when it is opposite particular places from particular locations, the location is warmed, so the movement kindles smoke from the dryness, and water vapor from heat."[21] Then, citing Aristotle or perhaps al-Kindi as "the wise one," *al-hakim,* Bayhaqi suggests how the universe came into existence through this smoke and vapor: "The Wise One maintained that these two vapors rising with the heat of the sun, when they meet in space, there the universe was possible and able to form. . . . They call the smoke *sulfur* and the vapor *mercury,* for when the universe came into being, an appropriate metal was brought forth from the essence of that place. . . . In the mines of the pounded stones, the most perfect and sublime of them is the sapphire, for it is the utmost end of nature. The remaining are gems such as emerald, garnet, diamond, turquoise, and others."[22]

In the *Javahirnama,* Ibn Mansur inquires into the physical shaping of rocks by noting the difference between things that simply exist (*vajib al-vujud*) and those that need a cause to exist (*mumkin al-vujud*). Rocks and stones (*hajjar*) belong to the latter category, being compounds (*murakab*) formed over years through natural processes and exposure to the elements. According to Ibn Mansur, this began with water and liquid matter (*jawhar-i abi*). As liquid matter came into contact with earth matter (*jawhar-i turabi*) and was exposed to heat and cold, it solidified, taking the shape of rocks and stones.[23]

In the mines, stones assumed their defining physical characteristic— their colors. The Persianate books of precious stones attribute the color and radiance of stones (*alvan-i javahir*) to their composition and the nature of the weathering process they undergo. These books also distinguish the formation of translucent stones from that of opaque ones. Translucent stones, such as rubies and diamonds, were thought to have retained more water during their formation, while opaque stones, such as turquoise and lapis lazuli, contained more earth. The spectrum of color ranged from dark (*bayaz*) to light (*savad*). According to traditions that Persianate books of precious stones relate, "the marriage of dark and light hues creates colors, and blending together, they bring even more colors into the world."[24] These cosmographies see the rays of the sun and stars as having dried and hardened earth matter into radiant and colorful stones. The *Javahirnama* of Ibn Mansur describes the known precious stones (*javahirat*) of the world in twenty chapters, with turquoise the subject of the eighth (see table 2), and has seven additional chapters on metals (*filizat*).[25]

Although the *Javahirnama* presents a geological classification of the formation of rocks, earths, and mineral substances, noting their uses and values, as Ibn Mansur's discourse on their supernatural powers

TABLE 2   THE TWENTY PRECIOUS STONES OF THE WORLD ACCORDING TO
MUHAMMAD IBN MANSUR'S *JAVAHIRNAMA*

Stone	Origin	Color
Pearl (*dur, marvarid, lulu*)	Islands of the Persian Gulf and the Indian Ocean, particularly Sri Lanka	White
Ruby and sapphire (*yaqut*)	Sri Lanka	Red, blue
Emerald (*zumurrud*)	Egypt and Sudan, particularly Aswan	Green
Topaz (*zabarjad*)	Egypt, particularly Alexandria	Green
Diamond (*almas*)	India	Transparent white
Cat's-eye (*'ayn al-hir*)	Sri Lanka	White and spotted like a cat's eye
Balas ruby (*la'l*)	Afghanistan, particularly Badakhshan	Rose
Turquoise (*firuza*)	Iran, particularly Nishapur	Sky blue
Bezoar (*pazahr*)	China, India	Yellow, green
Carnelian (*'aqiq*)	Yemen, particularly Sana'a, Aden	Red
Garnet (*banafsh*)	Afghanistan, particularly Badakhshan	Purple
Agate and onyx (*jaz'*)	Ethiopia, Yemen, China	White, black, red
Magnet (*maghnatis*)	Yemen, India	Iron
Emery (*sunbada*)	India	Red, black
Malachite (*dahna*)	Western Europe, Central Asia, Arabian Peninsula	Green
Lapis lazuli (*lajvard*)	Afghanistan, particularly Badakhshan	Azure
Coral (*marjan*)	Tunisia, particularly Marsa	Red
Jasper (*yashb*)	Yemen and China, particularly Kashghar	White, green
Crystal (*bulur*)	India, Central Asia, Arabian Peninsula	White
Amethyst (*jimast*)	Arabian Peninsula	Red, blue

reveals, early modern Islamic science still regarded natural substances as part of the manifold strange and wonderful forms of nature (*'aja'ib al-gharib*). This was in keeping with medieval Islamic natural histories, such as Zakariya ibn Muhammad al-Qazwini's compendium of the marvels of the universe, *'Aja'ib al-Makhluqat wa Ghara'ib al-Mawju- dat* (The wonders of creation and strange things existing). Completed in 1270, during the rule of the Mongol Ilkhanid dynasty (1256–1335), this work, which covered the wonders of the heavens as well as of the seas and the earth, accepted that while the natural world could be

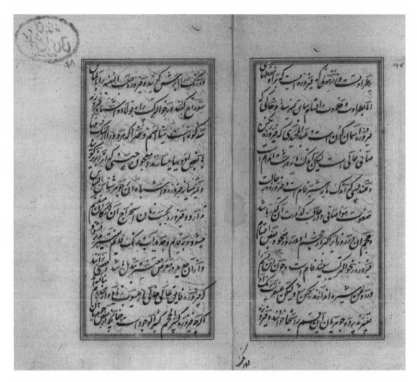

FIGURE 1. The *Javahirnama* (Book of precious stones) attributed to Muhammad ibn Mansur, on the qualities of sky-colored (*asuman gun*) turquoise. MS 2166 (1815), fols. 97–98, Majlis Library Archives, Tehran.

partially known through science, it also possessed unexplainable, even marvelous, qualities and patterns.[26]

Among the post-Timurid Perso-Islamic empires of the early modern Near East and Asia, turquoise was hallowed because of its celestial pale blue shade. Turquoise, stone and color, was called *firuza* in Persian and *fayruz* in Arabic, but its color (*rang*) was often referred to as *asuman gun* (sky colored) or *rang-i asumani* (the color of the sky). In his assessment of the turquoise trade, Ibn Mansur correlates the value of the stones with their luster and color (see figs. 1 and 2). "Turquoise is of different varieties," he begins, and "the one expert in knowing that jewel [*ma'rifat-i javahir*] knows as soon as he sees it what mine [*kan*] it originates from."[27] There were, according to the *Javahirnama*, five types (*qism*) of turquoise: Nishapuri, Ghaznavi, Ilaqi, Kirmani, and Khvarazmi. None had much value, however, and they were prone to lose their color—except for the turquoise found in the seven mines of "old rock"

FIGURE 2. The properties and prices of Persian turquoise. "Risala-yi Iskandariya va Javahirnama" (n.d.), attributed to Muhammad ibn Mansur, MS 5690/3, fols. 37–38, Majlis Library, Tehran.

outside the city of Nishapur, stones that were delicate and smooth (*latif u saf*) and of an unfading sky blue. The ultimate and maximum color (*bi qayat rangin*) belonged to the luminous turquoise of the old Abu Ishaqi mine. More varieties of sky-colored Nishapur turquoise came from the other old mines, which Ibn Mansur referred to as Azhari, Sulaymani, Zarhuni, Khaki, 'Abd al-Majidi, and Andalibi.[28]

According to Ibn Mansur, the color of turquoise was inconstant and determined by when it was taken from the mine (*zaman-i istikhraj az ma'dan*). The color of turquoise from the old mines was set, since it had already changed its color (*ab u rang-i qadim taqir pazir shavad*), whereas stones in the new mines were still changing their hue.[29] In the newer mines, large turquoises of lustrous blue could be found, but they soon faded to green. Color determined the value of turquoise on the market and was seen as the source of the stone's many virtues. Because of its color, turquoise was thought to have certain properties and uses (*khasiyat*), and popular belief esteemed the stone as a talisman, a sacred object that portended good fortune and could protect one from harm. The blue of turquoise, according to the *Javahirnama,* made it celestial earth, an amulet of victory, and a natural curative substance and medicine for the eyes.[30] Due to its cerulean shade, turquoise acquired significant meaning as sacred stone and color in court regalia and imperial cultures across Islamic Eurasia.

The differing grades of turquoise mined at Nishapur were sold by quality and size. The most valuable turquoises, with favorable size, shape, and color, were classified as ring stones, called *angushtari,* and sold by the piece. Smaller stones—sold by weight and often made into rings with more minute settings and other objects by jewelers—most commonly reached the markets of Central Asia, South Asia, the Near East, and Europe. In the fifteenth century, Ibn Mansur reported on the presence of abundant turquoise (*kasir al-vujud*) in Eurasian markets.[31] In Balad Sham, or Greater Syria, a flawless turquoise from Abu Ishaqi or Azhari weighing half a mithqal was worth seven to ten dinars, a one-mithqal piece twenty to thirty dinars, a two-mithqal piece fifty to seventy dinars, a three-mithqal stone one hundred to one hundred and fifty dinars, and so on.[32]

Turquoise was embedded in the culture of kingship and empire across Islamic Eurasia and known as a victorious stone adorning conquerors and kings. Ibn Mansur's *Javahirnama,* which covers the usual repertoire of turquoise's attributes but in greater detail, praises it as the stone of conquest (*hajr al-ghalaba*), the stone of honor (*hajr al-jah*), a jewel befitting Eurasian sultans, princes, and dynasts. It was believed that turquoise (*firuz*) brought victory (*piruz*), as the similarities in the two words suggest, and conferred the power to conquer enemies (*bar dushman zafar yabad*). Because turquoise was thought to be auspicious and good for the eyes, Ibn Mansur wrote, the rulers of old looked upon it when waking in the morning or on seeing the new moon (*mah-i naw*), making it the "stone of the eye" (*hajr al-'ayn*).[33] Eurasian princes and shahs made a great demand for turquoise and devised rituals around the mystical sky-blue stone. At the beginning of spring, in the festival of Nawruz, they found it pleasing to gaze upon precious stones, especially turquoise. On such occasions, Ibn Mansur recounts, royal cupbearers (*saqiyan*) would often place jewels in heavenly sherbets (*sharbat-i sarsabil*), which they served to the sultans.[34] Similarly, Bayhaqi's *Ma'din al-Nawadir fi Ma'rifat al-Jawahir* relates the virtues of turquoise in a purported letter on the art of kingship from Aristotle to Dhul-Qarnayn, the mythical Persianate Alexander:

It is known that turquoise is a stone the kings of Persia still adorn themselves with. Its quality is to drive away death from its wearer and bearer; it has never gleamed in the hand of a murderer, and what benefit is greater than this? He who looks upon the turquoise every morning, shall on this day be happy. . . . He who carries the turquoise shall triumph over enemies and shall be honored and respected in the sight of the people. For this, they call turquoise the stone of victory, the stone of honor, the stone of the eye. . . . It is among the attributes of Venus and God knows best.[35]

The Islamic genre of books of stones opens up and offers a point of entry into the world of meanings behind the Eurasian turquoise trade and its significance in interimperial exchange. As turquoise circulated throughout the Near East and Asia, so did knowledge of the culture surrounding it as a colorful and auspicious gemstone and an object of regalia symbolic of world-conquering empires and their courts.

TURQUOISE AND THE SAFAVID EMPIRE

In 1510, the city of Nishapur and its turquoise mines came under the control of the Safavid Empire, which eclipsed the Timurids in Khurasan. The Safavids emerged from a Sufi brotherhood in eastern Anatolia to subvert the White Sheep Turkmen dynasty and establish a Shi'i Islamic empire extending from Mesopotamia in the west across the Iranian plateau to Central Asia in the east. Shah Isma'il I (r. 1494–1524), the founding monarch of the empire, and his bands of mounted pastoral nomadic Turkic followers, the Qizilbash, followed the model of the Central Asian Timurids and their synthesis, moving from the tent to the throne to accept and become immersed in Perso-Islamic imperial culture. Along with the other post-Timurid empires of Islamic Eurasia—the Mughals, the Ottomans, and the khanates of Central Asia—the Safavids negotiated their sovereignty and power through tributary exchanges that included natural resources and objects such as precious stones.

For the Safavid Empire, such tributary exchanges were highly significant because money was scarce and had minimal purchase in early modern Iran. Although a trimetallic currency system of silver, gold, and copper coins—silver dirhams and golden dinars, expressed as fractions of the Mongolian tuman—had been in place in Iran and Central Asia since the times of the Ilkhanid dynasty in the fourteenth century, and major commercial business, such as the Safavid silk trade, involved monetary transactions, the bulk of everyday exchanges were not monetized, operating instead on the barter system and the trade and traffic of goods in kind.[36] Indeed, even the minting and use of coins were entwined in the customs and culture of imperial tribute, representing, along with the proclamation of the ruler's name in the mosques during Friday prayers (khutba), the establishment of sovereignty. Nor was the Safavid Empire particularly active in mining the deposits of precious metals throughout the country, relying instead on the inflow of silver from the New World via the Ottoman Empire.[37] The empire possessed abundant deposits of metals—including copper (mis) and iron (surb) in the mountains of Kirman, Mazandaran, and Yazd and steel (pulad) in Isfahan and Khuzistan—but mined them ambivalently.

Some distance from this bullionist and quantifying perspective on the mining of precious metals and stones yields a different view of the uses and meanings of material resources and objects across the tributary networks of the Safavid Empire, as the researches of James Allan on metalwork in the court culture of early modern Iran deftly show.[38] The other mineralia of the empire, precious stones of wondrous colors, were treasured: pearls (dur or marvarid) from the Indian Ocean and the Persian Gulf, balas rubies (la'l) and lapis lazuli (lajvard) from Badakhshan, and turquoise (firuza) from Nishapur—the richest mines in the empire.[39] These were greatly esteemed as gems and widely extracted from the earth and the sea under the auspices of the Safavid Empire. It was not copper itself but rather the colorful sedimentary flowers of copper, such as turquoise, that the Safavids sought and prized.

The turquoise mines on the outskirts of Nishapur held the most valuable mineral deposits in the empire. Miners followed established standards and methods for unearthing turquoise, digging galleries and tunnels through the rocks to allow air and light into the caves, their work overseen by engineers and the middlemen who transported the stones to the gem cutters of Mashhad. As it had been in the Timurid period, Safavid Nishapur was a provincial town, now increasingly linked economically and culturally to the Shi'i shrine city of Mashhad, about forty-five miles to the east. The pilgrimage routes to and from the tomb of the eighth Shi'i imam, 'Ali ibn Musa al-Riza, in Mashhad also became routes of the turquoise trade.

The Safavids placed the turquoise mines of Nishapur under direct imperial control. References in imperial histories such as Iskandar Bayg Munshi's Tarikh-i 'Alamara-yi 'Abbasi suggest that during the rule of Shah Tahmasp I (r. 1524–76), the Safavid state levied a 20 percent tax on the turquoise mines. These revenues were designated religious taxes (khums) and deposited in the state treasury (khazana-yi 'amira).[40] We read in Budaq Munshi Qazvini's *Kitab-i Javahir al-Akhbar*, completed in 1577, that the minister (*vazir*) of Khurasan acted as the prospector and overseer (*sarkar*) of the turquoise mines, keeping their accounts as part of the administration of the eastern province.[41] On occasion, such as in 1578, portions of the turquoise stones and dust (*khak-i firuza*) that had accrued in the royal treasury were distributed, by the order of Shah Tahmasp, among sayyids, the ulama, the poor, students, and people from every walk of life.[42]

Shah 'Abbas I (r. 1587–1629) put the mining and trade of turquoise from the valuable old rock—all its output—under a royal monopoly. This policy was in keeping with his mercantile tendencies and concerted efforts at establishing control over the empire, its resources, and its trade as a means of expanding the Safavid economy and its export revenues. The shah sought to stimulate the development of his mines, so much so that he reportedly ordered an unnamed European in his court "to inspect the mines all over Persia and see how they were being exploited."[43] He also issued ordinances restricting the export of raw and unset turquoise from the old mines, limiting work on the most valuable stones to Persian goldsmiths. Concerns that the old mines, of the finest turquoise, were becoming exhausted may also have spurred such careful measures to preserve these deposits for local craftsmen. Shah 'Abbas himself reserved the premium stones of the old rock, to store in his treasury and to present as gifts to kings and princes. What the sovereign did not keep, he sold and traded away, and it was through this means that stones of the old rock reached the markets of Asia and Europe.

In the Safavid period, new turquoise mines were opened to meet the global demand for the stone, but the specimens from the "new rock" were often inferior and could not match the unfading stones from the old rock. The seventeenth-century French gem trader and traveler Jean Baptiste Tavernier, famed for his "discovery" of the Golconda diamond mines in India and for selling the ill-fated 112-carat French Blue, also known as the Hope Diamond, to King Louis XIV of France in 1669, described the state of the turquoise trade in Safavid Persia. In 1632, he visited the turquoise mines of Nishapur and noted that the coveted stones of the old rock were under royal monopoly and could only be gifted by the shah, while stones from the recently opened mines of new rock could be bought and traded:

> Turquoise is only found in Persia, and is obtained in two mines. The one which is called 'the old rock' is three days' journey from Meshed towards the

north-west and near to a large town called Nichabourg; the other, which is called 'the new,' is five days' journey from it. Those of the new are of an inferior blue, tending to white, and are little esteemed, and one may purchase as many of them as he likes at small cost. But for many years the King of Persia has prohibited mining in the "old rock" for any but himself, because having no gold workers in the country besides those who work in thread, who are ignorant of the art of enameling on gold, and without knowledge of design and engraving, he uses for the decoration of swords, daggers, and other work, these turquoises of the old rock instead of enamel, which are cut and arranged in patterns like flowers and other figures which the jewelers make. This catches the eye and passes as a laborious work, but it is wanting in design.[44]

Most Persian turquoise stones that reached the world market were set in rings that were then strung together on a cord. By some accounts related in European lapidary manuals, this permitted prospective buyers to test the quality of many pieces at once, by seeing whether the stones lost their color when tucked under an armpit.

Evidence as to who the middlemen were in the turquoise trade is scarce, since European merchants and companies could not penetrate the trade. However, certain clues suggest that the Armenian merchants of the suburb of Julfa in the Safavid capital of Isfahan, with their long-distance trading networks, had a hand in the trade. In his seventeenth-century *Book of Histories*, the Armenian chronicler Arakel of Tabriz (c. 1590s–1670), quoting an Armenian priest in Aleppo named Kahana Sargis, details the attributes of turquoise largely as related in Persian books of precious stones, suggesting an Armenian link in the transmission of the stone and its culture across Asia and Europe. After identifying the varieties of turquoise and noting that the best stones came from Nishapur—the "Ishaqi" mine in particular—the *Book of Histories* offers variations on the properties of the stone: "Firuza . . . They say that whoever wears a turquoise stone on his finger will not have any need for money, and his words will be pleasing to the people. However, it does benefit one who wears a turquoise to utter unpleasant words, thinking that since he wears a turquoise his words are pleasant. The stone burns in fire."[45] In its qualities, turquoise exceeded all other precious stones: "Jewelers say that turquoise is better than all other gems, for it has many properties. Whoever wears it will not suffer the wrath of kings. If one looks at it in the morning, one shall not have any troubles in the coming day. The stone is beneficial for diseases of the eye. It brings long life, prosperity, riches, and helps against bad dreams."[46] *The Book of Histories*, reconfirming evidence provided in Jami's Timurid-era *javahirnama*, leaves little doubt that turquoise had reached European markets, giving its prices in European currencies. An Abu Ishaqi stone the size of an average kidney bean cost fifty reales, while "a good clear turquoise" of Nishapur weighing twenty carats garnered four hundred florins, and a clear one-carat stone fetched one florin.[47]

Under Shah 'Abbas II (1632–66) and his dynamic grand vizier Muhammad Bayg, the Safavid Empire renewed its efforts at prospecting for precious mineral deposits and created the office of *avaraja nivis-i ma'dan*.[48] In the following decades, the turquoise industry and its revenues became further immersed in Safavid administrative bureaucracy, as early eighteenth-century Persian manuals of government record. The late Safavid–era manual *Dastur al-Muluk* notes that it was the function of the *avaraja nivis-i ma'dan* to maintain registers of the revenues of all the empire's mines and to record them in a book of transactions.[49] This practice is confirmed by the eighteenth-century Safavid administrative manual *Tadhkirat al-Muluk*, which distinguishes the mines in the empire from crown lands (*khassa*) and counsels assigning an official called an *avaraja nivis* to keep their records.[50] As both of these manuals note, the turquoise trade was also closely tied to the office of the royal goldsmith, known as the *zargarbashi,* in charge of the royal jewelry workshop and its craftsmen, including lapidaries, stonecutters, jewelers, and enamelers.[51] Still, the existing source materials on the fiscal and mercantile aspects of the turquoise trade in the Safavid era are rather sketchy. It is better seen through the exploration of the tributary networks and interimperial exchanges among the Safavid, Mughal, and Ottoman Empires.

TURQUOISE IN THE TRIBUTARY ECONOMIES OF THE SAFAVID, MUGHAL, AND OTTOMAN EMPIRES

Turquoise evolved into an object of interimperial rivalry commonly exchanged in the vernacular economies of early modern Islamic empires, which knew it as the stone of "victory" (*piruza*) and "the stone of conquest" (*hajr al-ghalaba*).[52] Gifted as tribute and plundered as spoils of imperial campaigns, the blue stones of Nishapur spread through material and cultural interchanges among the Safavid, Mughal, and Ottoman Empires. Through such forms of economic and material redistribution, turquoise became a trade good of symbolic value in encounters between the royal courts of Islamic Eurasia.[53]

Between Safavids and Ottomans

Following the Battle of Chaldiran, in 1514, when the gunpowder-equipped armies of the Ottomans defeated the mounted Qizilbash archers of the Safavids, the former looted innumerable objects and artifacts, including turquoise stones, from the Hasht Bihisht, Shah Isma'il's palace in Tabriz, and took them to the Topkapi Sarayi in Istanbul. The battle was fought over the allegiance of the Turkic pastoralists in the Anatolian crossroads between the Safavid and Ottoman Empires. At the start of the sixteenth century, the victories of Shah Isma'il and his freewheeling bands of Qizilbash cavalry won converts to the messianic Shi'i ideology that the early Safavids espoused. Appealing in verse to the largely illiterate Turkic tribesmen in eastern Anatolia and claiming to be a representative of the hidden twelfth imam, Isma'il molded tribal and Sufi elements into a powerful religious movement and empire. During the Safavid period, Shi'ism became firmly established as the religion of a vast majority of the population of the Iranian plateau and served to set the land off from its Sunni neighbors, the Ottomans to the west and the Shaybanid Uzbeks to the east in Central Asia.

As Isma'il continued campaigning and converting among the Turkic pastoral societies of Anatolia, the Ottoman sultan Selim I (r. 1512–20) responded by marching an army of 140,000 men east, carrying a proclamation that Isma'il was a heretic. The Ottoman and Safavid forces met at Chaldiran in northwestern Azerbaijan in August of 1514. The Janissary standing army of the Ottomans, armed with gunpowder, overwhelmed the mounted Qizilbash archers, and according to the early sixteenth-century chronicle *Habib al-Siyar* by Khwandamir, Ottoman "cannons and matchlocks obscured the heavens with smoke."[54] The battle consolidated Ottoman control of the Anatolian frontier and shattered Isma'il's aura as a divine shah. With defeat clearly at hand, Isma'il and his retainers fled, followed by the disarrayed Safavid cavalry, numbering forty thousand men. The Ottoman army marched into Tabriz, occupying the Safavid capital for eight days before returning to Anatolia. With the Qizilbash increasingly undisciplined and fragmented, the Safavids could not control eastern Anatolia or the sacred cities of Iraq and were pushed back onto the Iranian plateau.

The Ottomans' brief occupation of Tabriz proved long enough for them to access the treasures that Isma'il had collected in the Hasht Bihisht palace. The opulence of the Safavid shah and his royal retinue, even the martial classes, was storied, and court chroniclers such as

Khwandamir described "horses with golden saddles and bejeweled reins, gilded swords with Egyptian scabbards, belts decorated with pearls, hats embroidered in gold, multicolored silk tunics, [and] innumerable coins and jewels" being distributed among the amirs, the commanders of the army.[55] A list in the Topkapi Sarayi archives of booty purportedly taken from the Safavid treasury includes an assortment of turquoise objects: iron war and parade helmets damascened in gold and studded with sky-blue turquoise (see plate 1); Qur'an manuscripts with turquoise-encrusted covers and bindings; metal flasks and bowls set with turquoise stones; swords, daggers, and shields inlaid with turquoise; chess pieces encrusted with sliced turquoise cloisons (firuzakari).[56]

It is likely that the Ottoman troops who plundered the shah's treasury retained many of its jewels and precious stones. Yet much of the loot was transferred to Istanbul and became part of the imperial treasury at Topkapi Sarayi. Tavernier, who was in contact with two former Ottoman treasury officials in India, described the abundance of these coffers (ambar), which were certain to have included not only stones from Chaldiran but also imperial regalia from diplomatic exchanges with the Safavids. The mineralogical objects that Tavernier noted included an abundance of turquoise, often inlaid in riding gear: "In the uppermost Coffer, are kept the Bridles, Breast-Pieces, Cruppers, and Stirrups, which are recommendable upon the score of the Diamonds, Rubies, Emeralds, and other Precious Stones, whereby they are enrich'd: but the greatest part of them is cover'd with *Turkish*-Stones, which they have the art of setting extremely well. It is a most astonishing sight to behold the quantity of those precious Harnesses."[57]

To memorialize his victory at Chaldiran and take measure of the worth of his plunder, Sultan Selim I commissioned a book of stones. The Edward G. Browne collection of Persian manuscripts at the University of Cambridge has a fragment of this lapidary text, titled *Risala dar Ma'rifat-i Javahir*, composed by Muhammad ibn al-Mubarak Qazvini.[58] The manuscript is divided into an introduction and two "mines" (ma'dan), the first containing twenty-one "caskets" (durj) of precious stones and the second eight "treasuries" (makhzan) of precious metals, but it breaks off in the middle of the sixteenth casket, on lapis lazuli, and has no colophon. The introductory passages include laudatory verses (bayt) in praise of Selim, comparing the Ottoman ruler, victorious in the aftermath of Chaldiran, to various precious stones and metals:

In the time of generosity, he was like a mountain of gold and silver—
kuh-i zar u nuqra vaqt-i bakhshish.
In the time of worship, he was like King David—*Davud sifat bi gah-i
parastish.*
He has inherited knowledge from ancestors—*miras girifta ilm az jad.*
He is like topaz in the eye of enemies—*dar chishm-i 'adu-yi din
zabarjad.*
His heart is the diamond of the people of the faith—*almas-i dil-i
'adi-yi din.*
He is the ruby in the heart of the poor and weak—*yaqut dil-i faqir u
miskin.*
A coral in the seas of exploration—*marjan dar bahar-i tahqiq.*
The agate mountains of Yemen exact—*kuh-i Yemen u 'aqiq tadqiq.*
Like a new sword unsheathed in war—*sayf-i jadid dar ma 'arakiha.*
The turquoise of famed mines—*firuza-yi kan-i ma 'rifatha.*
Victorious as turquoise in wars in every land—*firuz bi jang dar hama ja.*
The sultan of the world, Selim Sultan—*sultan-i jahan Selim Sultan.*
The Being of Eyes, and the Eye of Beings—*insan-i 'ayn u 'ayn-i insan.*[59]

With his book of jewels, Qazvini claimed to have compiled knowledge on the precious stones in the "bazaar of the world" (*bazaar-i
dunya*), bringing together that of "merchants of jewels" about value on
the global markets and that of "merchants of science" on medicinal and
therapeutic uses. He based its content, including the description of turquoise, almost exclusively on Ibn Mansur's *Javahirnama,* which he
deemed the best of the mineralogical treatises.[60]

Along with booty, the Ottomans brought various craftsmen to Istanbul from Tabriz in the aftermath of Chaldiran. With artisans from other
sites of Ottoman conquest—such as Cairo, Aleppo, and Damascus—
they worked the precious stones that adorned imperial regalia.[61] Among
their ranks in Ottoman workshops, as listed in a Topkapi register of
craftsmen from 1526, were stonecutters and engravers (*hakkakan*),
damasceners (*zirnishanan*), goldsmiths (*zargariyan*), and metalworkers
(*kaffatin*).[62]

The seventeenth-century Ottoman traveler Evliya Çelebi mentions the
crafts and trades associated with jewels and precious stones in his book
of travels, the *Siyahatnama.* Among the guilds and craftsmen in the
imperial city of Istanbul, one could find jewelers (*javahiriyan*), engravers
(*hakkakan*), and pearl stringers (*lulujiyan*). Some of these artisans prac-
ticed a technique known as *firuzakari,* layering sliced pieces of turquoise
on boxes, amulets, and other objects. There were also six hundred itiner-
ant merchants who sold diamonds from India, garnets from Ceylon,
opals from Ethiopia, coral from the Mediterranean, peridots from the

Red Sea, rubies from Badakhshan, and turquoise from Nishapur. "God only knows the extent of their riches and number of their jewels," Çelebi wrote.[63]

In the course of his travels, Çelebi also chronicled how turquoise turned up across the empire and was used and seen by ordinary Ottoman subjects. Describing the lofty minarets of the Süleymaniye Mosque, completed by the chief architect Sinan in 1557, he wrote, "The chief architect used the jewels to decorate the grooves of that minaret with artistry of all sorts, also for the marble roses decorating the inscriptions. This is why it is called the 'Jewelled Minaret'. Some of the stones shine when the sun's rays strike, but others have faded and lost their lustre from exposure to the heat, snow, and rain. However, in the centre of the arch above the qibla gate, there is a Nishapur turquoise as large as a round bowl that dazzles the beholder."[64]

In a section on "the guilds and professions existing in Istanbul, with the number of their shops, men, *shaykhs* and *pirs* [elders]," Çelebi describes the eight thousand predominantly Shiʻi water carriers who plied the neighborhoods of the city serving water in cups studded with precious stones: "They are all on foot [*saqa*], dressed in black leather jackets, carrying jacks on their backs; different ornaments of flowers made of wire are stuck in their heads, and in their hands they carry cups of crystal and china, the interior of which are shining with onyxes, jaspers, and turquoises, or golden tasses, out of which they give drink to Muslims, in remembrance of the martyrs of Kerbala, and wish health and prosperity to those to whom they administer the water, saying that they shall drink it to the health of Hassan and Hossein."[65] Reaching eastern Anatolia and the city of Tabriz in 1642, he observed a similar custom during the Ashura ceremony, commemorating the martyrdom of Imam Husayn, in the month of Muharram. "Water carriers dispense cold water and sweet sherbets, poured from their waterskins into bowls of crystal, even agate and turquoise," he wrote, adding that "the thirsty spectators drink these beverages in remembrance of the martyrs who suffered thirst in the plain of Yazid on the day of Ashura, intoning the verse: 'For the love of Husayn of Karbala, to health!'"[66] Elsewhere, Çelebi notes that disciples of the Gulshani Sufi order of Trebizond "wear necklaces of coral, jasper, and turquoise."[67]

Çelebi's travel narrative also leaves behind traces that turquoise reached Europe through the Ottoman Empire. Of the Stephan Monastery in Vienna, he wrote, "The walls and the small and large domes are covered inside and outside with mosaics of varicoloured precious stones and carved alabaster. The scoops in the moldings of the domes, walls and arches are adorned with glass beads and precious stones of various colours . . . . And the wall on the south side is decorated with valuable gems such as ruby, olivine, Nishapur turquoise, Yemenite carnelian, garnet, cat's eye, fish's eye, yellow sapphire, blue sapphire, amber, mother of pearl, and Ethiopian pearl."[68] At the annual fair at Doyran in the Balkans, "great merchants . . . lay out their wares—precious stuffs such as silks and satins and velvets; or jewels such as rubies, emeralds, chrysollite, turquoise, agate, etc.—turning their counter-tops into decked-out brides or idol temples."[69] Encounters and exchanges between the Safavid and Ottoman Empires occasioned the circulation of turquoise and its culture as a sacred stone and imperial jewel in the eastern Mediterranean world.

Indo-Persian Crossings

With the mines of Nishapur being located in the eastern Iranian province of Khurasan, which lay astride the trade routes to the Indian subcontinent, turquoise became an iconic object of exchange between the Safavid and Mughal Empires. The Mughals were a dynasty of Central Asian and Persianate Timurid origins that rose in northern India in the 1520s. The trade between the two empires thrived through the shared Persianate linguistic and cultural world that connected Mughal India and Safavid Iran. Traded across the Indo-Persian crossroads, turquoise—and its culture—was part of Mughal consciousness since the time of that empire's founding. In his memoirs of travel and conquest, *The Baburnama*, Zahir al-Din Babur reports on the turquoise mines in his native Ferghana, in Central Asia.[70] In the *Humayun Nama*, Babur's daughter Gulbadan Begum chronicles his gifts of jewels to his harem in Kabul following the Mughals' defeat of the armies of the sultan of Delhi Ibrahim Lodi at the Battle of Panipat in 1526. After "the treasuries of five kings fell into his hands," Babur ordered that "to each begam is to be delivered as follows: one special dancing-girl of the dancing-girls of Sultan Ibrahim, with one gold plate full of jewels—ruby and pearl, carnelian and diamond, emerald and turquoise, topaz and cat's eye."[71] These were the most valuable stones listed in the Persianate genre of *javahirnama*.

In 1529, Babur's son and heir apparent, Humayun, sponsored the composition of a book of precious stones to honor his father's empire. Dedicated to both of them, the *Javahirnama-yi Humayuni*, as it has come to be known, is a treatise by Muhammad ibn Ashraf al-Husayni al-Rustamdari, written, as the preface states, at a time when Babur had conquered northern India and possessed himself of the jewels that its previous rulers had stored in their treasuries.[72] Comprising twenty-two chapters (*bab*) subdivided into sections (*fasl*) on the qualities (*sifat*), mines (*kan*), prices (*qaymat*), and properties (*khasiyat*) of various stones, the text is an inventory of the mineralogical wealth of India, largely in the format of Muhammad ibn Mansur's *Javahirnama*.[73] The *Javahir-nama-yi Humayuni*, true to Timurid sensibilities, deems turquoise as having the best properties—an imperial stone favored by royals, such as King Solomon, who was said to possess a spacious turquoise throne.[74]

The premium turquoises that entered the Indian subcontinent were known to originate from the mines of Nishapur, in neighboring Safavid Iran. In 1544, Humayun—after having become emperor, been driven from the subcontinent by the Suri Afghans in 1540, and taken refuge in the court of the Safavid shah Tahmasp I—had visited the turquoise mines of Nishapur, which he viewed as among the sacred and marvelous wonders of creation, on his way to reconquer India. As the later Mughal history *The Akbarnama* describes it:

> On the 15th of Muharram, 951, he reached Holy Mashhad and visited the shrine of Imam Rizavi,—may the blessing of God be upon him! He spent some days in the presence of that noble building. Thence he went to Nishapur. . . . His Majesty visited the turquoise mines in that neighborhood, and from thence went to Sabzawar and from thence to Damaghan. Among the marvelous things of that place is an ancient fountain, which has a talisman from of old, to wit, whenever any dirty thing falls into the fountain a storm arises, and the sky grows dark from the force of the wind and the dust. This too he examined with the eye of prescience. How many things are there not in the wondrous workshop of the Creator, the understanding of which does not come within the scope of our thoughts and imaginings.[75]

Turquoise was a mineral marvel of the earth, an object of imperial fortune and victory, and thus became an amulet of friendship in contacts and exchanges between the Safavids and the Mughals, before they came into contention over the Afghan city of Qandahar. By the seventeenth century, the Mughals had incorporated turquoise in jeweled thrones displayed in elaborate court ceremonies marking the anniversaries of the coronation of emperors such as Jahangir (r. 1605–27) and Shah Jahan (r. 1627–58).[76] In the annual solar and lunar weighing ceremonies, Mughal emperors were counterbalanced against precious metals and stones, which they then dispersed among the princes in attendance.[77] In his diary entry for September 1, 1617, Sir Thomas Roe, the English envoy to the court of the Mughals, described such an event:

> Was the King's Birth-day, and the solemnitie of his weighing, to which I went, was carryed into a very large and beautiful Garden; the square within all water; on the sides flowers and trees; in the midst a Pinacle, where was prepared the scales, being hung in large tressels, and a cross beame plated on with Gold thinne, the scales of massie Gold, the borders set with small stones, Rubies and Turkey, the Chaines of Gold large and massie, but strengthened with silke Cords. . . . Suddenly hee entered into the scales, sate like a woman on his legges, and there was put in against him many bagges to fit his weight . . . with Gold and Jewels, and precious stones.[78]

Roe also remarked that the Portuguese relied predominantly on jewels in their trade with India: "We never heard of any commodity the Poertingalls doe bring to Goa then Jewells, ready mony, and some few other provisions of wine and the like, except the marfeel [ivory], gold, and amber they bring from Mozambique."[79] Many jewels and precious stones from India, Ceylon, Iran, and Afghanistan also reached Surat, farther north on India's west coast.[80]

The Portuguese had arrived in Goa a century earlier. Not long after, the Safavids had presented them and other South Asian rulers and potentates with gifts of turquoise. In 1514, seeking diplomatic ties with southern India, Shah Isma'il I sent offerings of the stone to Goa—including half an *alqueire,* or more than three quarts, "just in the same condition as when they came out of the mines" to the Portuguese admiral and viceroy of India Alphonse de Albuquerque and "a bowl of ordinary turquoise . . . a marvel to look upon," among other objects and stones, to the young Shi'i-leaning Isma'il Adil Shah of the Sultanate of Bijapur.[81] Shah Isma'il's gifts served as a reminder of the Safavids' presence in the Persian Gulf and their claims on the strategic island of Hormuz at a time when the Portuguese Estado had established itself as a major maritime power in the western Indian Ocean by seizing Goa in 1510 and gaining control of the pepper trade on the Malabar Coast. Fleets under Albuquerque's command had first appeared off Hormuz in 1507 and returned to take the island on behalf of the estado in 1515.[82]

Through India and Central Asia, turquoise also found a broad pan-Asian circulation as an object of exchange beyond the realms of Islamic empires. In Buddhist Tibet, where it was regarded as auspicious and therapeutic, the best stones were known to come up from India, while Buddhist art and iconography linked turquoise jewels to tribute-bearing peoples from the Near East, Central Asia, and the lands of Islam.[83]

According to one source, "a great merchant of Tibet who traded ages ago with India, and once crossed the seas beyond India, brought the finest real turquoise to his native country."[84] The traffic in stones between Tibet and India moved in both directions, with raw turquoise from Tibet being exported to India to be worked in gold and silver and then reimported into Tibet, along with prized turquoise stones from the mines of eastern Iran.[85] It is unclear how early turquoise came into use in Tibet, but the historical evidence suggests that the earliest fine turquoises to arrive there were Persian stones sent through India, where turquoise had spread by the fifteenth century.[86] In the record of his journey to Tibet, Osvaldo Roero notes that turquoise arrived there from Central Asia and lists Persian turquoise among the goods imported into Ladakh via Bukhara.[87] Further, the iconographic art of Lamaism represents the peoples of Islamic Eurasia making tributes and giving gifts to Buddha of precious stones, including turquoise, lapis lazuli, and rubies, suggesting their role in carrying this trade eastward to Tibet, where these gems came to adorn Buddhist idols and the "the whole range of Tibetan ethnography."[88]

Turquoise and its culture also spread through the Indo-Persian crossroads to the Turkmen, a conglomeration of mobile Turkish-speaking pastoralists inhabiting the steppes and oases of the Qara Qum, or Black Sands, desert, between the Caspian Sea and the Oxus. In 1576, as reported by Abu'l Ghazi Bahadur (1603–42), the khan of the Central Asian oasis of Khiva, the Oxus changed course, leading to the expansion of the sandy steppes of the Qara Qum.[89] Here Turkmen pastoralists forged a powerful and wide-reaching equestrian network through their domestication of swift Central Asian horses. Horsemanship had long lay at the foundations of the Turkmen pastoral economy. According to some accounts, Turkmen horses were mixed with Arab breeds, introduced into the Central Asian steppe at the turn of the fifteenth century in the reign of Timur Lang, who was thought to have distributed forty-two hundred of the best Arabian mares among the tribes.[90] Consequently, between the sixteenth and nineteenth centuries, the Turkmen carved out a loose trading and raiding confederation built on the power and speed of horses capable of making seemingly impossible journeys through the steppes.

The most renowned Turkmen horses were those of the Tekke people of the oases of Akhal. By most accounts, these breeds were beautiful and much sought after. Long, slender, and sinewy, they thrived on the dry grass of the desert and were well known for their fleetness and endurance on long marches through the steppes. Horseback riding was essential to the Turkmen's pastoral way of life, and the tribes were known for the great care they took of their horses, which were the basis of their mobility, independence, and power. This love of horses was not only practical but also ceremonial, and the Turkmen adorned their horses with felts, woven textiles, and ornaments of precious stones and metals on their heads, necks, bridles, and saddles.[91] Along with carnelian ('aqiq), the best of which came from Yemen, turquoise from the nearby mines of Khurasan was the stone that appeared most often in the material culture of the Turkmen. Sky-blue turquoise inlaid on the harnesses, collars, bridles, and saddles of Turkmen horses reveals the pastoral side of the Timurid Central Asian and Persianate synthesis that rendered the stone an auspicious talisman.[92]

Blues for Qandahar

Between the sixteenth and early eighteenth centuries, the oasis city of Qandahar in southeastern Afghanistan was on the caravan routes that connected Safavid Iran and Mughal India and traversed Central Asia. The city possessed a strong fortress and was surrounded with cultivable lands. Qandahar was of such importance to the Safavids and the Mughals that in the sixteenth and seventeenth centuries it was besieged fifteen times and changed hands on a dozen occasions, although it was only rarely taken by storm.[93] For the Safavids, the possession of Qandahar was essential for the preservation of Khurasan and its trade, while for the Mughals, the city was closely connected with the economy of Kabul, the base from which Babur had conquered his Indian empire. Inhabited predominantly by the Ghalzai tribes of the Pashtun confederacy, Qandahar was a vital strategic outpost between the Safavid and Mughal worlds, a prize that intensified imperial rivalries and strained the long-standing friendship between the two empires.

Circa 1500, Qandahar belonged to the Timurid dynasty of Herat. Following the Safavid defeat of the Timurids in Khurasan and conquest of Herat in 1507, however, Babur, of the rising Mughal Empire, took possession of the city and its fortress in 1522 as part of his conquering march from Central Asia into India and appointed his second son, Kamran Mirza, as its governor.[94] In 1537, Shah Tahmasp and thousands of his qurchi troops marched on Qandahar from Herat, briefly taking the fortress city before Kamran Mirza recovered it in 1538. In 1545, Humayun took possession of Qandahar with the help of Safavid troops, having earlier apologized to Shah Tahmasp, in whose court he took refuge, for his brother Kamran Mirza's stubborn refusal to give up the Afghan city.[95] Humayun handed the fort and its treasures over to Shah Tahmasp, but after most of the Persian expeditionary force had returned to Iran, Qandahar again fell into Mughal hands. Despite allusions to the fact that the fort of Qandahar remained a dependency of the shah and to Humayun's vow to restore it to Iran on reconquering India, the city was in effect part of the Mughal imperium. By the time of Humayun's death in 1556, his promise was still unredeemed, and with the ascension of the young Jalal al-Din Akbar to the Mughal throne, Shah Tahmasp renewed his campaign against the city, capturing it in 1558. The Mughals recovered it in 1595.[96]

The assertive new Persian dynast, Shah 'Abbas I, set out to retake Qandahar, regarding the fort city as rightfully belonging to the Safavid Empire. On securing the province of Khurasan and its most important oasis city, Herat, in 1598, Shah 'Abbas announced his victories and requested the return of Qandahar in a letter to Akbar. But his request went unheeded by the Mughal emperor, and as a result, retinues of the Safavid Qizilbash made forays against the city in the first decades of the seventeenth century. In 1607, early in Jahangir's reign, bands of Qizilbash from the eastern Iranian provinces of Farah, Herat, and Sistan independently laid siege to the city before being forced to retreat, at which point Shah 'Abbas suspended his attempts to recapture Qandahar by force and sought instead to win the friendship of, and thus soften, the newly enthroned Mughal emperor.[97]

Subsequently, the two kings exchanged letters referring to each other as brothers and proclaiming the unity of Indo-Persian territories between the Safavid and Mughal Empires as a shared and overlapping space. Their correspondence metaphorically expresses this sense of interconnection, a sort of "physical harmony." Following the New Year (Nawruz) in the spring of 1611, Yadgar 'Ali Sultan Talish, a Safavid ambassador to the Mughal court, presented Jahangir with rare gifts (including saddles set with turquoise stones) and a letter from 'Abbas praising the union and "proximity" of the Safavid and Mughal Empires across "physical and spatial distance."[98] Such declarations belied the contention between these empires over the possession of Qandahar.

The question of that fort city remained not too far below the surface and shaped the dynamics of Safavid-Mughal interactions and exchanges.[99] As the introduction describes, in 1613 the exchange of turquoise, the stone of victorious empires, became enmeshed in the imperial contention over Qandahar. In response to the request of Jahangir's jewel merchant Muhammad Husayn Khan Chelebi for turquoise, Shah 'Abbas informed him that the precious stone was under royal monopoly and could only be gifted by the king. Following this, Shah 'Abbas sent six bags containing a total of thirty seers of what he described as inferior turquoise to Jahangir, along with a letter continuing to profess his brotherhood and friendship. Although he reported that the gem was no longer mined as it had been, perhaps a reference to the depletion of the old rock, this may have been an intentional slight over the still outstanding fate of Qandahar. On receiving the bags, Jahangir noted with disappointment that "the royal gem carvers and setters did not find any of the stones worthy of being made into a ring."[100]

The poor quality of the turquoise that Shah 'Abbas sent, when he was in Mashhad, merely a few miles from Nishapur and the most esteemed mines of turquoise in the world, acquires fuller meaning when viewed in the context of the Qandahar question. Disgruntled that the Afghan fortress city had yet to be restored to the Safavid Empire, Shah 'Abbas may well have withheld the choice turquoise from Jahangir. The imperial contention and rivalry over Qandahar and the exchange of turquoise between the Safavid and the Mughal Empires provided the context for the miniature painting that Jahangir commissioned in 1618 from the court artist Abu'l Hasan representing his dominion over the Indian subcontinent and Central Asia. In the painting, Jahangir and Shah 'Abbas embrace while standing on a globe set against a turquoise sky (see plate 2). A larger-than-life Jahangir balances on a sleeping lion that straddles India and Central Asia, almost towering over Shah 'Abbas, who perches on a delicate lamb and is being pushed off the continent. The world-conquering Jahangir wears a ring of veined, sky-blue turquoise on his right hand. Here again, the outward appearance of brotherhood is undercut, by the oversize Jahangir's imperial sway over Eurasia and his possession of turquoise, the stone of victory and conquest.

The image of *Jahangir's Dream,* as the illustration has come to be known, could not be farther from the reality. In 1620, the determined Shah 'Abbas dispatched the embassy of Zayn al-Bayg to the Mughal court with the purpose of reopening the discussion on Qandahar. The embassy arrived bearing gifts, including a large ruby inscribed with the name of the Timurid prince, astronomer, and sultan Ulugh Bayg, to whose Central Asian heritage both the Mughals and the Safavids claimed an attachment.[101] But the Safavid embassy's references to the unsettled question of Qandahar were not met with a definitive response, thus shifting Safavid policies into a more aggressive mode. Having cultivated the confidence and trust of Jahangir, and with a garrison of merely three hundred Mughal troops guarding the fortress, the shah resolved to march into the Indo-Persian expanse and recapture Qandahar. In 1622, under the pretense of a hunting tour, the Qizilbash retook its fortress, "reduc[ing] the towers and ramparts to the ground and [bringing] the defenders to the point of amnesty."[102] Apart from a brief period when the Mughals recovered the city, from 1638 to 1649, Qandahar remained part of the Safavid Empire until its fall in 1722.

THE END OF THE TURQUOISE EMPIRE

From its distant point of origin in mines on the eastern marches of Iran, turquoise emerged into an emblematic object of interimperial exchange among the tributary empires of Central Asia, South Asia, and the Near East. The stone's trade and its culture peaked in the reign of the Safavid Empire but then fell fast. In 1722, a tribal rebellion originating in Qandahar, an object of Indo-Persian contention in the seventeenth century, toppled the Safavid Empire.[103] The Ghalzai Pashtuns of Qandahar—led by Shah Mahmud Hotaki (c. 1697–1725), the eldest son of an influential Ghalzai chieftain by the name of Mir Wais (1673–1715), who had resisted the Safavid yoke—marched on Isfahan, laid the Persian capital to siege, and vanquished the Safavid Empire, with ruinous consequences for the turquoise trade. With the state no longer in control of the turquoise industry, standards were forgotten and the most valuable mines, of the old rock, were abandoned (*matruk*) and fell into ruin. A century of dynastic instability followed, with no effort to revive the turquoise mines until the 1880s, in the reign of the Qajars. By then, the model of Islamic tributary empires in which turquoise had been ingrained as part of the projection and display of imperial fortune and power—the exercise of a symbolic, indirect, and layered sovereignty—had given way to new, more blunt and coercive forms of empire established by the colonial states and economies of the late nineteenth century.

CHAPTER 3

The Turquoise of Islam

As the turquoise trade spread in Timurid Central Asia, Safavid Iran, and Mughal India, the stone became converted into an imperial and sacred object projected in radiant displays across the blue-tiled cities of the eastern Islamic world. Its color marked the metropolitan architecture of oasis cities along the trade routes of Islamic Eurasia, visible from Timurid Samarqand and Herat to Tabriz of the White and the Black Sheep dynasties and from Safavid Isfahan, built in the time of Shah 'Abbas I, to the Friday Mosque of Shah Jahan at Thatta in the Mughal province of Sindh. Tile makers and ceramicists applied the shades of turquoise and azure to glazed tiles fired in kilns. Mixing roasted copper, lead, and tin, craftsmen produced the turquoise color of *firuza*. From cobalt and roasted copper came the night-blue azure known as *lajvard*. Set on the surfaces of mosques, madrassas, and mausoleums, glazed polychrome tesserae fired in the blues of Islam connected imperial cities and urban spaces across the Timurid, Safavid, and Mughal imperiums.

In recent years, the fields of Islamic, Near Eastern, and South Asian art history have seen the appearance of a blossoming literature on the imperial construction of urban space, introducing new approaches to the older historiography focused on the groundwork of cataloguing Islamic architecture. The works of Gulru Necipoglu, Shirine Hamadeh, and Zeynep Çelik on the Ottoman Empire; Kishwar Rizvi and Sussan Babaie on Safavid Iran; and Thomas Metcalf, Catherine Asher, Stephen

Blake, and Alka Patel on Mughal India, among others, have explored the role of the built environment in the construction and display of imperial sovereignty and power across the cities of the Near East and South Asia, bringing fresh perspectives to the study of such subjects as everyday life in urban spaces; the aesthetics of dynastic shrines, mosques, gardens, and palaces; and colonial visions of the Islamic architectural landscape.[1]

The following pages turn to this theme of architecture and imperial power by exploring the effects of the Eurasian turquoise trade on the geography and the spatial economy of Timurid, Safavid, and Mughal cities. Starting in the late fourteenth century, as Turkic pastoralists settled down, accepted Perso-Islamic culture, and rebuilt imperial cities integrated into worldwide networks of trade, the turquoise trade spread, and blue became the most visible color of architectural monuments in the eastern Islamic world. The seven celestial shades of the *haft rang*—turquoise blue, night blue, black, green, red, ocher, and white, all rooted in the mystical romances of the twelfth-century Persian-language poet Nizami Ghanjavi—came to color the oasis cities of early modern Eurasian empires.[2] Most of all, cobalt oxides and copper ores decorated the polychrome faience of imperial monuments with dark and light blues. The turquoise city, with its cerulean domes, minarets, and arches, built between sand and sky, became the expression of the widespread material culture of Timurid Central Asia and the Persianate world.[3]

These imperial cities lay astride the historic overland trade routes that linked together the steppes of the Oxus (Amu Darya), India, and Iran. In guiding the flow of merchants, pilgrims, and travelers, this chain of trade routes and oasis cities facilitated the spread of goods and commodities through the caravanserais and bazaars of Eurasia. Such trade goods included the opaque blue precious stones of Khurasan, the turquoise of Nishapur and the lapis lazuli of Badakhshan, as well as the pigmented earths and mineral ores—cobalt oxides and copper ores—that supplied the substance necessary for the blues of Islamic imperial architecture. The use of blue as the celestial shade and unifying color of imperial architecture in Central Asia, Iran, and India closely paralleled the expansion of the turquoise trade. A great number of these monuments were mosques and shrines and thus, as places of pilgrimage (*ziyarat*), contributed to the expansion of traffic and trade along the crossroads of Eurasian empires. This architecture was visible to the pilgrims and merchants who passed to and from these cities.

TURQUOISE THRONES

Turquoise became ingrained in the imperial cultures of early modern Eurasian dynasties that moved from the tent to the throne, a measure of power, conquest, and kingship among Turkic steppe peoples as they accepted and re-created the customs of Persianate culture. Along with building monumental cities, these empires strove to re-create the legendary Turquoise Throne (*takht-i firuza*), one of the bejeweled royal seats of the ancient Persian kings described in the epic *Shahnama,* or *Book of Kings,* compiled by Abu'l Qasim Firdawsi. Another was the famed Peacock Throne (*takht-i tavus*), said to have been in use from the reign of the legendary Iranian king Jamsheed through the times of the Achaemenid Empire (550–330 B.C.E.), until Alexander of Macedon broke it down upon the Greek conquest of Persia and sent it in pieces to Rum (the lands of the Roman Empire in the eastern Mediterranean).[4] Seeking the revival of the Achaemenid imperial standard, Ardashir Babakan (d. 242 C.E.), the founder of the Sasanian Empire (224–651 C.E.), searched for the lost Peacock Throne, but to no avail. According to legend, in subsequent years more than a thousand skilled craftsmen from China, Makran, Baghdad, and Fars came together to rebuild this throne, along with the other Persian bejeweled seats of power. By the reign of Khusraw Parviz (570–628 C.E.), these thrones had reached their full splendor. The gem-studded Peacock Throne—adorned with gold, silver, and 140 turquoises set in patterns—became the paramount seat of Sasanian imperial power. The three others were called the Ram's Head, bedecked with its namesake object; Lajivard, or Azure, which wind and dust had never touched; and Turquoise, which warmed the hearts of all onlookers.[5]

Turquoise thrones appear in various episodes of the *Shahnama,* which describes the seats of power of the "Kayanid Crown" as studded with the stone. In one story, Firaydun seeks to leave behind such a throne for his sons.[6] On the completion of his seven trials, the epic hero Rustam reposes on "a throne set with turquoise and ornamented in ram's horns," given to him by King Kavus.[7] After crossing into Turan, Rustam comes to rest on a turquoise throne in Turkistan.[8] Sitting beside a turquoise throne, the moon-faced Sudabeh tries to seduce her stepson, the tragic hero Siyavash.[9]

The turquoise throne also entered South Asian imperial cultures, as the rulers of the Deccan Sultanate aspired to ascend to the coveted *takht-i firuza.* According to Richard Eaton, the famed Turquoise Throne of the Deccan was crafted by Telegu artisans . . . framed in ebony, covered with plates of pure gold, studded with precious gems, and enameled with a turquoise hue. It had originally been built for Sultan Muhammad bin Tughluq, probably intended as tribute during the period that the Tughluqs ruled Sultanpur/Warangal (1323–36). But the Tughluqs were driven out of Warangal before the throne could be delivered to Delhi. Ultimately, Kapaya Nayaka transferred it to Sultan Muhammad I (1358–75) as part of a treaty agreement in which the two rulers fixed their common border. It was then used by every Bahmani ruler until the last, Sultan Mahmud (1482–1518), who dismantled it for its valuable gems.[10]

This throne became one of the most potent symbols of Bahmani power in the Deccan in the reign of Sultan Firuz (r. 1397–1422), a contemporary of Timur and the builder of a turquoise city known as Firuzabad.[11]

THE COLORS OF STONES

In the early modern period, Eurasian empires began to envision and construct their monumental capital cities in the ethereal hues of turquoise thrones. From the oases of Samarqand and Herat in Timurid Central Asia to Tabriz and Isfahan in Safavid Iran and to the cities of Thatta and Hyderabad in the Mughal province of Sindh, Islamic empires commissioned monumental complexes and urban spaces clad in turquoise and azure faience and glazed tiles (*kashi*). Skilled craftsmen sketched geometric patterns and inscribed words, prayers, and poems in mosaics of blues.

The sources of these imperial blues were colored earths, the elements and compounds used to pigment ceramic tiles. Abu'l Qasim 'Abdallah Kashani's fourteenth-century book of stones, *'Ara'is al-Javahir va Nafa'is al-Ata'ib* (Statements on jewels and gifts of rarities), details the art of ceramics (*kashi gari*).[12] According to Abu'l Qasim, it is rooted in the knowledge of raw material substances, namely stones (in Arabic, *hajjar*; in Persian, *sang*), and the colors that can be derived from them. From *shikar-i sang*, a white stone that looks like lumps of sugar, came transparent milky tints. Another white stone, called *batanih*, produced the color silver. From *muzarrad*, a kohl-like stone, came black. From oxides of lead, referred to as *madarsang*, came red. From cobalt, mined in the village of Qamsar in the mountains around Kashan, came a deep blue with the same name as the mineral, *lajvard*. Coated and glazed in this color, tiles were referred to as *lajvardina* ware. And from roasted copper, called *nuhas-i muharraq*, and the lead known as *usrub*, tile makers produced the color turquoise.[13] Abu'l Qasim describes how clay and earthenware tiles coated in turquoise glaze, after being fired in the kiln for twelve hours and allowed to cool for a week, came out an opaque sky blue. In some instances, when they came out of the firing white, they were painted with the enamel of "pure turquoise."[14] For Abu'l Qasim and his contemporaries, there was little doubt that colors derived from stones and the earth's substances.

These colored earths were employed on a wide scale in Timurid imperial architecture in Central Asia, providing the model for a dynastic metropolitan style that spread throughout the Near East and South Asia. By the seventeenth century, blue ceramic tiles adorned the walls, arches, minarets, and domes of Friday mosques (known as *masjid-i jami'*), madrassas, mausoleums, and Sufi shrines and lodges in urban spaces across the eastern Islamic world. Signature features of this new metropolitan style etched the skyline of cities in blues: the domed chamber (in Persian, *gunbad;* in Arabic, *qubba*), a lofty and geometrical interior space that was a microcosm of the world; the minaret (*minar*), from which sounded the call to prayer; the arched gateway (*pishtaq*) and the barrel-vaulted chamber (*iwan*), which led inside the structures; and the intricate honeycomb corbel vault (*muqarnas*) that pleased the eye.[15] Architects (*mi'mar*), builders (*banna'*), ceramicists and tile workers (*kashi tarash*), painters (*naqqash*), calligraphers (*katib*), and other trusted workers (*kargaran*) toiled for years to construct blue mosques and monuments that came to epitomize the display of imperial power.[16]

BLUE CITIES

The architecture of the blue city first materialized in the Timurid Empire. A world conqueror of Turkic pastoral nomadic origins, its founder, Timur, claimed descent from the Mongols and was keen to present himself as a rightful Perso-Islamic ruler on his ascent to the throne. The monumental imperial architecture of Shahr-i Sabz and Samarqand, which Timur chose as his capital in 1370, projected this image. Subsequently, in the fifteenth century, Timur's son and successor Shah Rukh (r. 1405–47) and Shah Rukh's queen Gawhar Shad continued these metropolitan architectural projects across the Oxus, in the city of Herat and its surroundings.

The travel narrative and chronicle of Kamal al-Din 'Abd al-Razzaq Samarqandi, a mid-fifteenth-century ambassador from the court of Shah Rukh to the South Indian kingdom of Vijayanagara, surveys the geography of the Timurids' turquoise cities. The account of his journeys from 1442 to 1445, *Matla' al-Sa'dayn va Majma' al-Bahrayn* (The rising of the auspicious twin stars and the confluence of the ocean), also covers the history of the cities and monuments of the Timurid realm until 1470.[17] It chronicles Timur's early campaigns and his first capital, called Shahr-i Sabz, or Green City, even though its predominant shade was blue, as seen on such monuments as the world conqueror's Aq Saray Palace, an edifice that the imperial chronicler Sharaf al-Din 'Ali Yazdi describes with awe in the *Zafarnama,* or "Book of Victories," completed circa 1436:

> The loftiness of its pinnacles reached such a height that the very sky fell from the eyes of the stars.
> In the dark night, the white light of its walls made the muezzin think it was time for the morning call to prayer. . . .
> Because its surface was the color turquoise [*rang-i firuza*], it became one with the stars.[18]

But 'Abd al-Razzaq and his contemporaries reserved their greatest praise for Samarqand. 'Abd al-Razzaq recounts the famed history of the imperial city's grand Friday Mosque, also known as the Mosque of Bibi Khanum, after Timur's Chinese wife, who, legend says, had it built while the conqueror was away, although royal chronicles suggest that Timur ordered its erection on his return from the conquest of Delhi in 1398. The mosque was to be one of the grandest and loftiest in the world:

> In the times when the Lord of the Auspicious Conjunction of the Planets [*Hazrat-i Sahib Qiran*] was campaigning against Hindustan, occupied with destroying the sanctuaries of the idol worshipers and the fire temples of unbelievers, he resolved to return to the capital of Samarqand and to build a Friday masjid in that unparalleled city. And when the retinues of the World Conqueror [*Mavakib-i Jahangusha*] reached the capital in the holy month of Ramadan, when all the customs of the times were carried out in the name of increasing worship and prayer, his Majesty ordered the construction of the Friday Mosque.[19]

The most talented and trusted workers—collected from the distant corners of the empire—toiled away at the task: "Skilled engineers and unparalleled architects laid the foundations of the mosque, while gifted craftsmen with dexterous hands," including numerous stone gravers from the regions of Azerbaijan, Fars, and Hindustan, "carried out the plans with precision." According to 'Abd al-Razzaq, even newly encountered fauna and beasts of burden were enlisted in the labor, as "95 chained elephants sent from India to Samarqand were used to draw enormous stones" necessary to construct the grand mosque.[20] An illustration by the Timurid court miniaturist Kamal al-Din Bihzad of the construction of the Friday Mosque appears in a manuscript of the *Zafarnama*.[21]

'Abd al-Razzaq praised the colorful earth substances that went into the construction of the Friday Mosque in verse, evoking the value that the tributary economy and imperial culture of the Timurid dynasts placed on these stones, a value greater than that of the most precious metal—gold:

> They build monuments from stones
> Stones that bring colors into the world.
> Don't think that the stones of the mosque are from gold
> Stones themselves are gems more valuable than gold.[22]

Their blues were displayed in the shrines and mausoleums of the Shah-i Zinda (Living king) cemetery and on the fluted dome of the Gur-i Amir (Grave of the prince)—a building of marble painted in lapis lazuli— where Timur and his sons were buried.[23] Timur's successors continued his imperial style in Samarqand, most prominently in the Ulugh Bayg Madrassa, built by his grandson in 1420, which flanked the main square of the Rigistan (Place of sand), the city's commercial center. The monumental, blue-tiled architecture of Samarqand continued to develop into the seventeenth century, under the patronage of the successors to the Timurids, the Shaybanid Uzbek amirs, who also transplanted the Timurid metropolitan style into the famed oasis city of Bukhara. The gold-covered Tala Kari madrassa was constructed in 1660, completing the Rigistan ensemble in Samarqand, while the Shir Dar, or Lion's Gate, madrassa, featuring on each side of its turquoise entrance portal two roaring lions fronting the sun, in reference to the Persian monarchical symbol *shir u khurshid*, was completed in 1636.

In the reign of Shah Rukh and Gawhar Shad, the Timurid metropolitan style flourished in the cities of Khurasan, as the urban centers and shrine cities of Herat and Mashhad were expanded and rebuilt in turquoise and night blue. In the north of Shah Rukh's capital of Herat, the Friday Mosque and the madrassa of Gawhar Shad were constructed in the suburban complex that became known as the Musalla, and farther north of the city, the shrine of the Sufi saint Khwaja 'Abdallah Ansari was rebuilt, while in Mashhad, the Friday Mosque was raised inside the shrine complex of the eighth Shi'i imam, 'Ali ibn Musa al-Riza (765–818). These monuments, among many others, were the work of the architect Ustad Qavam al-Din, a talented engineer, designer, and craftsman brought from the southern Persian city of Shiraz, possibly in the campaigns of Timur, to Khurasan, where he rose to influence in the court of Shah Rukh. Qavam al-Din was the master architect of the blue mosaic faience and polychrome *haft rang* tile work that brilliantly ornamented the buildings and skylines of Timurid cities in Khurasan.[24]

In 1424, Shah Rukh enclosed the tomb of the eleventh-century Sufi saint Khwaja 'Abdallah Ansari, raising a high arching turquoise portal and an *iwan* covered in turquoise-glazed tile epigraphy and geometric patterns at the shrine, about three miles north of Herat's city gates in a cemetery known as the Gazargah. 'Abd al-Razzaq recounts the craftsmanship that went into this construction: "In the foundation of that sublime monument, after the stones and bricks were set with limitless skill and finesse, it was surfaced with glazed tiles of melted gold and the azure of *lajvard*, adorned with the lines of Mongol, Kufic, and Persian scripts." The effect, in his words, was that the lofty blue shrine "became one with the sky."[25] It also led to an increase of traffic to the saint's tomb, as evident from its appearance in the pilgrimage manuals of the times, including the account written circa 1460 by the pilgrim known by the name of Sayyid Asil al-Din Va'iz Haravi.[26] In his detailed description of the Gazargah, just as in his survey of Imam Riza's shrine in Mashhad, 'Abd al-Razzaq links the imperial construction of such monumental shrines (*'amarat-i mazar*) to the expansion of networks of pilgrimage (*ziyarat*).[27]

All that remains of the Musalla ensemble are the ruins of a domed mausoleum and two minarets in turquoise blue—wreckage of colonial empires and their "great games."[28] In the fifteenth century, the ribbed cupolas and more than twenty minarets of this complex, the masterpiece of Gawhar Shad's architectural projects, saturated the city with turquoise tiles.[29] 'Abd al-Razzaq memorialized their construction: "In this year [1433], the sublime madrassa being built on the order of Gawhar Shad north of the city at the head of the Injil Bridge was completed. This building is such that it has no likeness in all the world. . . . It would take an eloquent scribe from the court of the highest sphere, with a pen made of the alloy of jewels and melted gold of the sun on a slate of silver luminous like the moon, to write its description." In verse, he praised the madrassa's radiant colors: "It became the aurora of mercury [*shafaq-i sangraf*] and the celestial sphere of azure [*gardun-i lajvard*]."[30]

Two hundred miles to the east, in Mashhad, the Masjid-i Jami'—the Friday Mosque constructed in 1418 on the order of Gawhar Shad and thus known locally as Masjid-i Gawhar Shad—welcomed crowds of pilgrims on their journeys to the shrine of Imam Riza.[31] "All that needs to be said is "Ali ibn Musa al-Riza' and nothing more [*Ali ibn Musa al-Riza guyi u bas*]," 'Abd al-Razzaq wrote in his chronicle in reference to the monument to the martyred Shi'i saint, which was built to expand pilgrimage to the shrine.[32] On reaching the shrine city, pilgrims saw the cobalt-blue tesserae of the mosque, with inscriptions of the word *Allah* etched on its surfaces.[33] Etchings on the turquoise- and cobalt-tinted tiles of the mosque, built through the patronage of a Sunni empire for a Shi'i population, repeated the Shi'i Muslim *shahada* "There is no God but Allah, Muhammad is the Messenger, 'Ali is the friend of God [*La ilaha illa Allah, Muhammadan rasul Allah, 'Ali vali Allah*]."[34] Still other inscriptions memorialized Gawhar Shad and the famed Qavam al-Din, the respective patron and architect of the mosque.[35] It is worth noting that the shrine town of Mashhad, located only one hundred miles from the turquoise mines, henceforth became a hub of Shi'i pilgrimage and Khurasan's trade. The construction of the Masjid-i Gawhar Shad near the shrine of Imam Riza, it could be suggested, spurred the pilgrimages of countless Shi'i devotees and merchants and prompted the expansion of the turquoise trade of Khurasan. Networks of shrines and pilgrimage across the empire augmented the trade and export of commodities and precious objects such as turquoise.

TABRIZ AND THE TURQUOISE OF ISLAM

From Samarqand and Herat, the Timurid Empire's metropolitan style of turquoise architecture spread westward to Azerbaijan, the territory of its rivals, the Turkmen Qara Quyunlu and Aq Quyunlu, centered on the city of Tabriz. In 1465, a mosque complex was constructed in Tabriz of tiles so blue that it came to be known as Firuza-yi Islam—the Turquoise of Islam. Built on the order of the Qara Quyunlu sultan Abu'l Muzaffar Jahan Shah and his wife Khatun Jan Begum, both of whom were laid in its mausoleum, the Turquoise of Islam was saturated with a mosaic of colors—most of all, blues created from cobalt and copper ores. From a distance, travelers to and from the city could see the azure and turquoise of the mosque, the tiles of its lofty domes and minarets blending color and form and seeking oneness with the sky.[36] The mosque, part of a complex including a bazaar, a madrassa, gardens, shrines, and mausoleums, remained in use under the Qara Quyunlu's successors in Tabriz, the Aq Quyunlu and Safavid dynasties.

Writing a description of the city in the early sixteenth century, Michele Membré, a Venetian traveler and envoy to the court of Shah Tahmasp I, praised its Blue Mosque: "At the entrance to the city of Tabriz from Anatolia it is all gardens and mosques with blue vaults. . . . There is a mosque with two minarets, which are like tall bell towers. This mosque is so well built that neither in the land of the Turk, nor in the lands I have seen, have I found another such building. Outside is all designs of foliage like porcelains, and with beautiful colored marbles."[37]

Positioned along the eastern Anatolian frontier, the mosque was caught in Safavid and Ottoman border wars and looted by victorious Ottoman troops following the Battle of Chaldiran in 1514.[38] Repeated earthquakes between the sixteenth and eighteenth centuries compounded the damage of these episodes of warfare.[39] By the seventeenth century, when the French traveler Jean Baptiste Tavernier visited the mosque, it was already in ruins, degraded because, as he reported, it was built by Sunni Turkmen and frequented by "Hereticks":

> There are to be seen at *Tauris* [Tabriz], Ruines of stately Edifices round about the great *Piazza,* and the neighboring Parts; they have also let run four or five Mosquees of a prodigious height and bigness. The most magnificent and the biggest stands as you go out of the Town. The *Persians* will not come near it, but look upon it as defil'd, and a Mosque of Hereticks, in regard it was built by the *Sounnis,* or the followers of Omar. 'Tis a vast Structure fairly built. . . . It is lin'd with brickwork varnish'd with different Colours; and adorn'd within with very fine painting *A Pantique,* and abundance of Cifers and *Arabian* Letters in Gold and Azure.[40]

With its blue domes and minarets, the ruined mosque captivated Tavernier:

> Upon two sides of the Fore-front are rear'd two *Minarets* or Towers very high . . . lined with varnish'd brickwork . . . and at the top stand two *Cupolas.* . . . Entering the Door of the Mosquee, you come into a spacious *Duomo,* thirty six Paces in Diameter. . . . The inside of the Walls is varnish'd in Squares of several Colours, with Flowers, Cifers, and *Arabian* Letters intermix'd, and wrought in Emboss'd-work, so well painted, so well gilded, that it seems to be but one piece of Work, cut out with a pair of Scissars. From this *Duomo* you pass to another lesser, but more beautiful of its kind. . . . The inside of the vault is a violet Enamel, painted with all sorts of Flowers in Flat-work, but the outside of both the *Duomo's* is cover'd with varnish'd Brick-work, and Flowers emboss'd *A la Moresque* . . . which diversity of Colours is very pleasing to the eye.[41]

The mosque was also known in its time as Muzaffariya Gunbad, in honor of the Qara Quyunlu ruler who commissioned it.[42] By the nine-

FIGURE 3. The Turquoise of Islam: the ruined Blue Mosque of Tabriz. Eugène Flandin and Pascal Coste, *Voyage en Perse: Perse Moderne*, vol. 8 (Paris: Gide and J. Baudry, 1851), plate 5.

teenth century, however, having fallen into a noticeable state of devastation, it was referred to as Masjid-i Kabud—the Blue (or Bruised) Mosque (see fig. 3). It was in such a state of decay that in the late nineteenth-century geographical history of Tabriz *Tarikh va Jughrafiya-yi Dar al-Saltana-yi Tabriz,* Nadir Mirza predicted that it would soon vanish, lamenting that "the day will come when eyes will no longer be able to gaze upon such an intricately magnificent edifice."[43] He further praised the monumental blue structure, writing that there was no way to do it justice by describing it or even to fully envision all that was to be seen in the place—including its colors. Still, he tried to put its blue remains into words: "Its turquoise-hued tiles seemed more beautifully colored than real turquoise stones."[44]

THE PATTERN OF THE WORLD

The imperial architecture of the turquoise city culminated with the construction of Safavid Isfahan. Shah ʿAbbas I favored this city in central Iran for hunting trips and made it the seat of his empire during an excursion in 1598. According to the court historian Iskandar Bayg Munshi in *Tarikh-i 'Alamara-yi 'Abbasi,* the shah brought together master architects and engineers to plan a city of immense scale, consisting of monumental buildings tiled in blue and gold:

> In the spring of 1598, he approved plans for the construction of magnificent buildings in the Naqsh-i Jahan district, and architects and engineers strove to complete them. From the Darb-i Dawlat, which is the name for the city gate located in the Naqsh-i Jahan precincts, he constructed an avenue to the Zayanda Rud. Four parks were laid out on each side of the avenue, and fine buildings adorned each. The avenue was continued across the river as far as the mountains bounding Isfahan to the south. The emirs and officers of the state were charged with the creation of the parks, each to consist of reception rooms, covered ways, porticos, balconies, finely adorned belvederes, and murals in gold and lapis lazuli.[45]

They arrayed an intricate collection of mosques, madrassas, bazaars, caravanserais, and palaces in blue ceramic tile around the square of the Maydan-i Naqsh-i Jahan (Pattern of the world) at the center of the city and along the tree-lined Chahar Bagh Avenue that led to the Zayanda Rud River and the thirty-three-arched Bridge of Allahverdi Khan.[46]

Such a radical reclamation of nature was not uncharacteristic of 'Abbas's efforts to make the oasis of Isfahan thrive. Among other projects, he attempted to divert the flow of the winding Karun River through the rocky and unsurpassable Zagros Mountains toward the Zayanda Rud, ordering the local Bakhtiyari pastoral nomads to dig a tunnel for it through the cliffs. But it was the rebuilding of his turquoise-tiled capital and its urban spaces in the metropolitan style of the Timurid imperial city that stands as his most successful project of environmental transformation.

The color turquoise was most brilliantly displayed on the ceramic-tiled domes of the newly constructed monumental mosques on Naqsh-i Jahan Square (see plate 3). In 1611, Shah 'Abbas ordered the construction of a Friday mosque of immense scale, a monumental project that by some accounts took nearly two decades to complete, on the southern end of the Maydan-i Naqsh-i Jahan, across the Qaysariya Portal of the bazaar. Made of the finest marble from a newly discovered quarry and surfaced with turquoise- and cobalt-colored tiles, the Royal Mosque stood as the crowning spectacle of his Isfahan (see plates 4 and 5). As recorded in *Tarikh-i 'Alamara-yi 'Abbasi,* the mosque was planned on the grandest scale:

This year, Shah 'Abbas conceived the idea of building a great mosque adjacent to the Naqsh-i Jahan Square in Isfahan—a mosque which would be without equal in Iran and possibly in the entire civilized world. Shah 'Abbas had made the city of Isfahan like a paradise with charming buildings, parks in which the perfume of the flowers uplifted the spirit, and streams and gardens. He had already built a mosque and a theological seminary on the eastern and northern sides of the square, respectively, but he was dissatisfied with these. Just as the city of Isfahan was the envy of other cities in respect of its residential buildings, its Qaysariya, its *caravanserais,* and its markets, and called to mind the Koranic verses: "Eram, possessor of lofty buildings, the like of which have not been created in these regions," the Shah wanted its mosques, seminaries, and pious foundations also to be the finest of their kind in Iran, and to rival the temple at Mecca and the mosque at Jerusalem. . . . At a propitious hour determined by the astrologers from an examination of the royal horoscope, skilled architects and engineers, each of whom claimed to be without peer in the science of engineering, laid the foundations of the mosque and sanctuary and the master craftsmen set to work.[47]

The ethereal, sky-blue mosque was completed in 1630, a year after Shah 'Abbas's death, becoming a landmark monument in the city of Isfahan. Flanking the eastern side of the square, the Shaykh Lutfallah Mosque, with its subtle arabesque and geometrical patterns in turquoise, night blue, and yellow, was built in 1618.[48]

Although never again on the same scale as in the first decades of the seventeenth century, the urban development of Isfahan continued under the successors to Shah 'Abbas, who embellished the city until the last days of their dynasty: the cerulean monument Madrassa-yi Madar-i Shah (College of the mother of the shah) and its adjacent caravanserai were built on Chahar Bagh in 1714, just eight years before the fall of the Safavid capital to the Ghalzai Afghans (see plate 6).

TURQUOISE CITIES OF SINDH

The Timurid metropolitan style of turquoise cities with blue-tiled imperial architecture reached Mughal India with certain variations and was integrated into its capitals. Amid the red sandstone and white marble monuments of Delhi, Fatipur Sikri, and Agra, subtle shades of Timurid blues revealed a Central Asian imperial connection. In the southern gateway to Akbar's Tomb in Sikandra near Agra, completed between 1612 and 1614, the red sandstone of the structure contrasts with the ornate marble inlay panels executed in blues and other lighter tones. Decorated with the inscriptions of 'Abd al-Haqq Shirazi, later known as Amanat Khan, this gateway was built as a South Asian variation of

Timurid imperial architecture.[49] Also recalling the Timurid imperial vision, the Taj Mahal in Agra, completed in 1653 on the order of the Mughal emperor Shah Jahan to enshrine his favorite wife, Mumtaz Mahal, is adorned with a bounty of precious stones from across Asia's expanse—including sky-colored turquoise from Iran and Tibet, night-blue lapis lazuli from Afghanistan, verdant jade from Kashghar, golden amber from Burma, and flamelike coral from the Indian Ocean. Through the method known as *pietra dura*, called *pachikari* in Agra, small slivers of these precious stones were arranged into intricate botanical patterns embedded in the marble surface of the Taj Mahal, which, like the tomb of Akbar, is also adorned with the epigraphic inscriptions of Amanat Khan.

It was in the oasis cities of the province of Sindh, west of the Indus River and inland from the Arabian Sea, however, that the Mughals constructed their bluest monuments in the Timurid metropolitan style. There the Mughal Empire found strong links to the brick and blue-tiled architecture of Central Asia and Iran, including an existing array of Timurid-era blue-toned architecture to build on. In Multan, Mughal sultans and princes came face-to-face with the white dome and blue tile–banded designs of the tomb of Rukn-i Alam, constructed under the Tughluq sultans of Delhi in 1320. In Uchh-i Sharif, a remote oasis in the southern Punjab, the Mughals encountered the blue-and-white-tiled crested tomb of Bibi Jawindi, a thirteenth-century female Sufi saint, built in 1498.

The Mughal response to and interpretation of the blue cities of Sindh was most boldly displayed in the Jami' Masjid of Shah Jahan in the city of Thatta. Built in 1647, at a time when Shah Jahan had occupied Balkh and was pressing toward Samarqand, this mosque further introduced architectural designs from the oasis cities of Central Asia into India. In the early seventeenth century, Shah Jahan had proved his attachment to the regalia of Persianate kingship by commissioning a replica of the legendary Peacock Throne as the seat of imperial power in Delhi, where it remained until Nadir Shah Afshar carried it off after conquering the city in 1739. Shah Jahan ordered the Friday Mosque in Thatta to be built in return for the hospitality he had received in Sindh while seeking refuge from his father, Jahangir. Its construction was likely part of the rebuilding of the city after monsoons ravaged it in the summer of 1637. The imperial chronicles of Shah Jahan's reign, appearing in the genre of *padshahnama* (book of the emperor), report that Thatta was nearly destroyed:

Throughout the city and all the neighboring *parganas* [districts] that bordered on the sea, rain had fallen unceasingly for some 36 hours with such violence as to remind one of the universal deluge. Most of the buildings were overthrown by this terrific hurricane of wind and rain; and large numbers of men, horses, and all sorts of cattle perished from it. The force of the gale . . . tore up stout trees by the roots and lashed the waves of the sea into such a tempest that countless myriads of fish were cast upon the beach. About 1,000 ships . . . were either wrecked in the open sea or stranded on the coast. These disasters caused heavy losses to all concerned; but in addition to them, the sea, through the force of wind and water, on this occasion rose so considerably as to flood a large tract of country, which was impregnated with the salt held in solution by the water and thereby rendered incapable of cultivation.[50]

The Shah Jahan Mosque in Thatta was built during the city's reconstruction after the 1637 storm. With its large courtyard and elaborate corridor of ninety-three domed chambers, this mosque has the most elaborate display of tile work in the Indian subcontinent. Its domes, arches, gateways, and vaulted interiors are inlaid with mosaics of radiating turquoise and white tiles in floral patterns that recall the *kashi* work of Timurid Central Asia and Safavid Iran.[51] Mughal experimentation with turquoise-tiled architecture continued into the eighteenth century, with the construction of the tombs of the amirs of Sindh in Hyderabad (see plates 7 and 8).

IMPERIAL CITIES

Between the fifteenth and seventeenth centuries, the blue architectural monuments of the Timurid imperial cities of Samarqand and Herat became manifold across the eastern Islamic world, from Tabriz of the Black Sheep and the White Sheep to Safavid Isfahan to the Mughal province of Sindh. Imperial cities were rebuilt in the turquoise of Islam, a color that came from stones and earth substances that spread along the trade and pilgrimage routes connecting Central Asia, South Asia, and the Near East. The expansion of the turquoise trade paralleled the emergence of the built environment of the turquoise city—the visible display of monumentality and power across the urban landscapes of early modern Islamic empires.

Turquoise became the shade of imperial cities where sovereignty and command were expressed and negotiated in part through monumental architecture and vivid color displays. Through the construction of ethereal mosques, shrines, and madrassas adorned with polychrome glazed tesserae, the architecture of Islam transformed into projections of imperial power and dominance. As Kamal al-Din ʿAbd al-Razzaq Samarqandi repeated in variations of a verse when describing the Timurids' monumental spaces of pilgrimage and veneration in his chronicle and travel account:

> The best work of shahs while on the throne
> Is to solidify the foundations of the faith
> By building monuments of stones,
> Stones that bring colors into the world.[52]

Rebuilt and arrayed in shades of blue, the architecture of oasis cities across the Timurid, Safavid, and Mughal Empires spurred the expansion of traffic along the routes of pilgrimage and trade. The construction of turquoise-tiled cities paralleled the height of the Eurasian turquoise trade. Stones of turquoise, mined in the mountains of Nishapur, at the crossroads of the overland routes linking Central Asia, Iran, and India, were transported, strewn, and scattered, carrying with them their material culture and meaning as imperial and sacred objects.

Stone from the East

From the early modern empires of Islamic Eurasia, turquoise reached Europe, becoming coveted and collected as a rare and peregrine stone from the distant East. The stones that arrived in the markets of Europe came overland from Iran through the Ottoman Empire—and were thus known as "Turkish," *turchese* in Italian and *turquoise* in French. European travelers encountered turquoise following the long-distance voyages of the continent's seaborne commercial empires to Asia in search of commodities and places to establish trading colonies. These empires and their commercial economies arrived on the heels of what is commonly referred to as the age of exploration and "discovery," when they endeavored to capture the trade with the Indies and, along the way, came across and conquered the Americas.[1] First the Portuguese Estado, then the Dutch East India Company (Verenigde Oost-Indische Compagnie; VOC) and the British East India Company (EIC), forged commercial empires through the global trade of major profitable commodities—primarily spices, textiles, silver, and gold. Along the way, early modern European travelers, merchants, and naturalists happened upon and collected a wide array of material objects, as well as their meanings. These objects of exchange included new flora, fauna, and mineral substances, such as precious stones, that expanding empires and their subjects strove to collect and classify.[2]

Turquoise and the other colorful precious stones of the world became the subject of early modern European natural histories and geological texts. The encounter with the stones and ores found and traded across the globe—and the need to know their origins, properties, and values— precipitated the advancement of the earth sciences. Between the fifteenth and the nineteenth century, an unprecedented volume of European natural histories classified and made legible the mineral world that lay below the surface of the earth, tracing its physical features. Merchants and naturalists identified the gamut of jewels and what was known about them, incorporating indigenous knowledge on stones, including material from the genre of *javahirnama*, into their accounts.

Turquoise appeared in Europe and in this literature as a strange and unfamiliar stone, a colorful exotic and talisman from faraway lands. It became known and was sought after for its singular, unusual sky-blue shade and was one of the colorful natural substances that gave matter to the chromatic spectrum of the world, spurring the construction of a diversity of hues.[3] In particular, azure-tinted stones from the lands of Islam—turquoise and cobalt from Iran; lapis lazuli from Afghanistan— introduced the color and the phenomenon of blue, an unstable and once uncommon shade, across Europe.[4] But turquoise and its culture were ultimately read differently in Europe, in whose commercial economy they found a different purchase, than they had been among the Islamic tributary empires of early modern Eurasia. As stones were classified by hardness and translucence and as their value was monetized, turquoise was demystified and lost its meaning, now deemed superstition, as a sacred stone and an expression of imperial power.

AZURES

The construction of blue materialized with the export of colorful flora and *mineralia* from Asia. Lapis lazuli from Afghanistan became the source of the esteemed blue *ultramarino*, treasured by the painters of the Italian Renaissance, while Persian cobalt was used prolifically as the blue pigment in Asian porcelain and tile. The shade of urban architecture in the imperial capitals of Islamic Eurasia imitated the color of the turquoise stone. As the global trade of lapis lazuli, cobalt, and turquoise spread stones in shades of blue around the world, the meanings they possessed went with them.

In *The Pilgrim Art: Cultures of Porcelain in World History*, Robert Finlay explores the construction of "Muslim blue" in Eurasian cross-cultural exchanges. China used this term for the cobalt that it imported from Iran for the coloring of porcelain that it then traded across the world, binding the color blue and the world of Islamic Eurasia together.[5] Taking this notion of the provenance of blue as a point of departure, it is possible to trace how the trade of lapis lazuli and turquoise stones gave matter to the perception and the construction of the colors azure and sky blue, respectively.

The first blue arrived from Asia in the form of lapis lazuli and cobalt. The former is a pigment and a hard stone found at the Sar-i Sang mines near Badakhshan in the mountains of northeastern Afghanistan, where it is called in Persian *sang-i lajvard*. The word *lajvard* may also apply to cobalt ores mined near the Persian city of Kashan and traditionally used as pigments of deep blue resembling lapis lazuli.[6] The Arabic derivation of the name, *lazward*, is the root of one of the most common words denoting the color blue: *azure*.[7] The ancient Egyptians knew lapis lazuli as a precious stone of deep blue, veined with gold, that could be used as a natural blue pigment. The now-destroyed Buddhas of Bamiyan, near the mines of Badakhshan, were once surfaced in lapis. Among the post-Timurid empires of Eurasia, *lajvard* was one of the *haft rang*, the seven colors of heaven, prominently featured in the glazed tiles that adorned Islamic architectural monuments. The finest blue pigment was obtained from lapis lazuli.

During the Renaissance, the Italians increasingly demanded the beautiful deep blue that lapis lazuli created. The artists of the period called this color *ultramarino*, to denote that its source came from far away, "beyond the sea."[8] The stone was rare and expensive because it was an exotic object that had to be brought over a long distance, from the mountains of Afghanistan, and its hardness made extracting it from the earth difficult. The process of turning the stone into a pigment, through pulverization and purification, was also slow and complicated by the necessity of removing the nonblue particles. Lapis lazuli creates a diverse range of deep blues as a pigment, but its density means it does not spread easily over surfaces, and for this reason, along with its cost, the striking ultramarino was usually kept to small areas of paintings.[9]

In Renaissance Florence, this blue was deemed the only color worthy of the Virgin Mary's robe.[10] Because there were cheaper substitutes available, such as German blue, Italian painters specified their desire for ultramarine when contracting their work. For his *Adoration of the Magi*, the Florentine artist Domenico Ghirlandaio even particularized the quality he sought: "The blue must be ultramarine of the value of about four florins per ounce."[11] He demanded ultramarine because it was the premium and best shade of blue in the world.[12]

The import of dyeing materials from India and the Near East introduced rich new colors to Renaissance Italy. By the late fifteenth century, the demand for exotic hues had given rise to shops specializing in pigments and related materials.[13] Strolling through these, Renaissance artists would have found various substances from the Indies and the Near East: lapis lazuli, brazilwood, kermes, saffron, fenugreek, orpiment, and indigo. All of these pigments would have been procured from Venice, where merchants brought the dyeing materials they imported from the East.[14] Indigo, an exotic plant dye imported from the Indies, was another important source of blue.[15] This rich colorant came from the leaves of the bushes of the *Indigofera* genus, which grew in tropical India. Indigo could be found in Venice by the twelfth century, and by early modern times, Portuguese, British, and Dutch traders were producing it in their colonial botanical ventures. These imports readily found a market, as indigo yielded a blue that was far deeper and darker than that of woad. By the late seventeenth century, the exotic dye plant had supplanted woad and colored the world in Indian blue. The cultivation and trade of the plant in India, and later in the colonies of the New World, created an unprecedented supply of and demand for blue tones in Europe throughout the seventeenth and eighteenth centuries.[16] *Indigo,* like *lapis lazuli* (and its cognates *lajvard, lazward,* and *azure*), came to refer to a shade of blue, with Sir Isaac Newton designating it one of the seven colors of the rainbow in 1660.[17]

Turquoise, a phosphate of aluminum and copper mined in eastern Iran, was another Asian commodity of chromatic importance, and as it traveled around the world it took on meaning as the color of the sky.

JOURNEYS TO THE TURQUOISE EMPIRE

In early modern Europe, turquoise became associated with tales of fantastic voyages and encounters in the East. Among the first to describe the stone was Ruy González de Clavijo (d. 1412), an ambassador from Henry III of Castile to the court of Timur Lang in Samarqand, who passed through Nishapur in 1404, describing it as "a great city . . . very large, and supplied with all things."[18] The nearby turquoise mines were still in operation, yielding the world's premium stones. "Here they find turquoises [*turquesas*]," Clavijo wrote, "and, though they are met with in other places, those of Nishapur are the finest that are to be come by."[19]

According to Clavijo, these turquoises—along with pearls from the Persian Gulf and balas rubies from Badakhshan in the mountains of

Afghanistan—constituted the mineral wealth of the Timurid realm and were prominently displayed at court: "Standing and set beside this table, was to be noticed a golden tree . . . bearing leaves like oak trees. The fruit of this tree consisted in vast numbers of balas rubies, emeralds, turquoises, sapphires and common rubies with many great round pearls of wonderful orient and beauty. . . . The rubies, turquoises, and other gems grew from the branches."[20] To meet the demand for turquoise, "day by day men go and seek and break into the rocks on that mountainside to find these precious stones. When the vein is discovered where they lie, this vein is carefully followed, and when the jewel is reached it must be cut out little by little with chisels until all the matrix has been removed." Jewelers then cut and polished the stones by "grinding the gem on millstones." So valuable were the turquoise mines for the empire that, as Clavijo reported, "by order of Timur a strong guard had been established at the mines to see to it that his Highness's rights were respected."[21]

On the Anatolian crossroads, the Timurids' western rivals the Turkmen Aq Quyunlu (White Sheep) were middlemen in the overland trade between Asia and the Mediterranean. Trade links between the Aq Quyunlu capital of Tabriz and Venice spurred the transmission of turquoise and other precious stones from Asia to Europe. The Venetian merchant and explorer Joseph Barbaro, a visitor to the court in Tabriz in 1474, remarked on the city's position as a thriving entrepôt for "jewelles, sylkes . . . and spices . . . with verie great trafficque of merchaundize."[22] Elsewhere, Barbaro reports on the wide array of Asian precious stones that the tributary economy of the White Sheep dynasty trafficked to the court of Uzun Hasan in Tabriz, including a collection of large balas rubies of "good color" and "great perles and turcasses" of old.[23]

In the sixteenth century, the Safavids, the successors of the Timurid and Turkmen empires, further opened Iran to the world, and an increasing stream of Europeans traveled to the country and wrote accounts of their journeys there. Some of the earliest European accounts of turquoise in Safavid Persia depict its use in royal circles.[24] Michele Membré, a sixteenth-century Venetian traveler and dragoman to the court of Shah Tahmasp I, noted the protocol and sartorial custom that surrounded turquoise among the *qurchis,* royal guards selected from the Qizilbash tribes:

> All the king's *qurchis* wear swords with scabbards of massy gold—that is those *qurchis* who have done some deed of bravery—and the dagger, which they call *khanjar,* of massy gold with turquoises, and also all the Lords. And

they wear a belt of massy gold with turquoises and rubies. . . . They also wear on their heads, upon their turbans, a strip of gold with turquoise and ruby stones. And there are those of them who wear three or four gold strips; and they wear plumes, and at the foot of the said plumes there is massy gold with precious stones. And when the King makes any festival, they all dress like this.[25]

Over time, these travel narratives began to survey Iran's natural resources and identified the mines of turquoise in the district of Nishapur as valuable for trade.[26] These mines also appear on Dutch maps of Safavid Iran, including *Persia sive Sophorum Regnum* (Persia or the kingdom of the sophy), by Willem Janszoon Blaeu, the official cartographer of the VOC, printed in Amsterdam in 1635 as part of the world atlas *Theatrum Orbis Terrarum*. Blaeu's hand-colored map, which depicts Safavid Persia at the height of its power, extending from Baghdad to Herat, includes an image of Shah ʿAbbas I flanked by two of his guards and notes the location of the mines in Latin: "Hic turchini gemme repriuntur" (Here turquoise gems may be discovered).[27]

In 1660, the Capuchin missionary Raphaël Du Mans noted the presence of a lively "*trafic de turquoises (phérousé [firuza])*" in his description of the trade and commerce of the empire.[28] In 1674, the Venetian Ambrosio Bembo visited the Maydan-i Naqsh-i Jahan in Isfahan, where he discovered "a whole bazaar of jewels where one can find many turquoise and lapis lazuli stones that are gathered in the provinces."[29] In his travel account, Bembo depicts a flourishing trade in precious stones among the Safavid, Ottoman, and Mughal Empires, along overland and Indian Ocean routes. Turquoise from Iran had already reached India and come to be regarded as having medicinal properties by the early sixteenth century. John Fryer, a doctor in the service of the EIC in Surat who wrote extensively on the natural history of the Indo-Persian world, remarked on the Safavid Empire's most precious stone: "Deeper in the Bowels of the Mines, the Turquoise (the most lively of any) endures the rape of those that search for it."[30]

The voyages of the French jeweler, merchant, and journeyman extraordinaire Jean Chardin (1643–1713), however, led to the most significant by far of the early modern travel accounts of Safavid Persia and the most detailed descriptions of the empire's turquoise. The son of a Huguenot jeweler, Chardin first journeyed to Asia on behalf of his father in 1665 in search of diamonds. Traveling through Isfahan on his way to Mughal India, he gained access to the court of Shah ʿAbbas II (r. 1642–66), who appointed him a royal merchant and commissioned him to purchase jewels and pattern jewelry based on the shah's own designs. In Paris in 1670, Chardin completed this task, but when he returned to Isfahan in 1673, bearing "jewels of goldsmith's work and precious stones visibly fine," he found that the shah had passed away and been replaced by his ineffectual son, Sulayman (r. 1666–94). What is more, the prices of precious stones had fallen and the tastes of the new court were austere. In short, the royal superintendent Najaf Quli Bayg informed Chardin that Shah Sulayman did not covet jewels and that "the times of the late king had passed and were over."[31] The head of the goldsmiths (zargarbashi) also noted the empire's decay of commerce since the death of Shah ʿAbbas II, which had, among other things, halved the value of precious stones.[32] Seeking to recoup the expenses incurred in seven years of journeys and in need of money to purchase diamonds in the Indies, Chardin rejected Najaf Quli Bayg's offers to compensate him in kind with turquoise, silk, and diamonds. They ultimately settled on a price of one thousand five hundred tumans (the equivalent of seven thousand Spanish pistoles or seventy thousand French crowns), which amounted to less than half the gains Chardin had expected but allowed him to make an end of this transaction of gems with the Persian crown.[33]

As might be expected, Chardin's account of his travels, the monumental *Voyages en Perse,* surveys in detail the mining and trade of metals and precious stones in the Safavid Empire. It notes that the mountainous land of Iran was rich in minerals and that Shah ʿAbbas I had worked to develop mining in the early seventeenth century, including opening up new mines:

> Since Persia is very mountainous, it is full of metals and minerals, which have begun to be pulled from the ground in this century, much more than in preceding centuries. It is Shah Abbas the Great to whom this is owed. The great number of minerals found throughout the kingdom compelled him to work the mines. The metals found most widespread in Persia are iron, steel, copper, and lead. One finds no gold or silver. It is impossible for mountains that produce all sorts of metals, and sulfur and saltpeter, to not also produce minerals of the sun and moon.[34]

Chardin was quick to add, however, his opinion that these resources were still underutilized, writing, "The Persians are lazy to make many discoveries. They do not investigate further than what they have always had and do not seek to acquire anything more. If they were as active, as concerned, and as needful as us, there would not be a single mountain peak which had not been dug many times."[35]

In the third volume of *Voyages en Perse,* Chardin gives an intriguing account of the empire's turquoise mines and the trade of its stones:

> The richest mine of Persia is that of turquoises. One finds it in two places, in Nishapur in Khurasan and from a mountain four days' journey from the Caspian called Firuz Kuh or the Mountain of Firuz, who was one of the ancient kings of Persia. . . . The mine of turquoises was also discovered during the reign of Firuz and took its name from him. So does the fine stone that we get out from it, and call turquoise, for in the East they call it Firuz. But the best is that of Nishapur. One pulls from those mountains turquoise of an incomparable beauty and bounty. One exploits them easily, excavating by the well method. Sometimes one discovers veins of turquoise and finds invaluable jewels. . . . They have since discovered another mine of these sort of stones, but they are not so fine and lively. They call them the new rock to distinguish them from the others which are old: the color of it goes off in time. All of the turquoise from the old rock is kept for the king, who selects the most beautiful and alive, with the remainder being trucked away and traded. The miners and customs officials are apt to overlook the king's monopoly and divert what they can and from there it comes to pass that one so often has a good chance to find these old rocks and turquoise.[36]

Despite the royal monopoly on turquoise from the old rock, fine stones that had been picked over and refused by the sovereign or smuggled by miners reached the world market, along with the turquoise of the new rock, which the shah's subjects could readily mine and legally trade.

Given access to the royal treasury in Isfahan, Chardin saw gems "from all parts of the world" ("toutes les parties du Monde"), including emeralds from South America and rubies from Asia, and admitted that he "would never have believed there could be so many jewels and riches."[37] He describes heaps of turquoise stones ("des monceaux de turquoises") amassed like a harvest in the Safavid treasury: "One sees a chamber full of turquoise; the stones, rough and uncut, are piled high on the floor like heaps of grain, and the polished stones fill innumerable big sacs of leather, weighing forty-five to fifty pounds each."[38] Not only this abundance but also its neglect amazed Chardin: "It is not that surprising that the king possesses such a treasure of turquoises, [because] the mine is in his empire; but what I find extremely shocking is that one lets such riches be turned to dust and broken to pieces while heaped on top of one another."[39]

Chardin also detailed the revenues and tributes that the Safavids derived from turquoise. He noted their methods of collecting taxes on turquoise mines not under royal monopoly, such as those of new rock: "That which one pulls from the mines, like the mines of turquoise, the mines of copper, and others, is on the condition that one pay for the expenses of the work and put back the value of twenty *mescals* of gold, weight by law (two and a half ounces); some mujtahids, excluding this last condition, say one must give one-fifth of what one benefits from."[40] On other occasions, Chardin observed the lavish and extravagant uses to which the Safavids put the stones, such as encrusting swords ("épées de turquoises") then valued at four hundred pistoles, which the court gave as gifts of honor (*khil'at*) to envoys from European companies.[41] In his travels through the Safavid realm in the seventeenth century, Chardin encountered a culture of imperial regalia studded with turquoise stones.

THE BLUEST STONE

In its journeys across the world, turquoise carried its meanings along. Its culture arrived in Europe, with some variations, in early modern times, where the stone was one of a number of colorful exotic commodities from Asia, Africa, and the New World. European printed texts from that era on mining and metallurgy feature turquoise in some detail, classifying it physically as a mineral substance and culturally as a shade of the color spectrum. In both cases, its sky blue identified and defined it.

The publication of hundreds of volumes on rocks, metals, and minerals advanced the earth sciences in early modern Europe. These often appeared in the lapidary genre, about the medicinal and alchemical properties of stones. Scientific forms of knowledge surpassed without completely leaving behind older belief systems about the powers of stones.[42] Lapidary manuals on the earth's jewels and mineral substances combined geological sciences with persistent notions of the astral and magical qualities of stones. For instance, after identifying and naming turquoise and its color, European naturalists considered its various miraculous virtues and supernatural properties: medicine for the eyes, reflection of its owner's health or the fidelity of a lover, remedy for scorpion bites, and talisman against falling off horses.

An early reference to the stone appears in a lapidary text published in Lyon circa 1495 and attributed to John Mandeville, the mythical traveler whose printed tales of fabulous journeys and adventures introduced Asia, Africa, the Near East, and the Indian Ocean to European readers.[43] Titled *Le Lapidaire en Francoys,* the text claims to be the translation for René d'Anjou of a Latin source and to contain information on "the properties and virtues of various gems; the names and colors of distinct stones and the places they are found."[44] It has a highly superstitious, at times foreboding, description of the strange and wonderful qualities of the mysterious turquoise: "Turquoise is a milky green stone from the Orient. More than all other stones, it strengthens the vision and guards against adverse causes and accidents; it gives boldness and grace to those that wear it, and horses sweating and hot from exertion will not be chilled from drinking cold water. The people of India and the Orient say that if the stones are attached to the bridles of horses, it protects the animals during battle and guards man from death. . . . It possesses such power that the man who owns it cannot beget offspring and the woman cannot conceive."[45]

More popular works of literature also mention turquoise's mysterious appeal. William Shakespeare's *The Merchant of Venice*, completed in 1598, alludes to the stone's reputation as a prized, exotic commodity and luxury from the East. In the third act, the merchant Shylock is inconsolable on discovering that his cherished turquoise, a gift from a past love, is gone, traded for a monkey by his daughter: "Out upon her: thou tortur'st me *Tuball*, it was my Turkies, I had it of *Leah* when I was a Batchellor: I would not have given it for a wilderness of Monkies."[46] Shylock mourned this loss because turquoise was a valuable Oriental commodity, an object of unknown virtues, as well as, in this case, a keepsake from a past love.

One of the earliest European mineralogical accounts of turquoise appears in *Speculum Lapidum* (1502) by Camillus Leonardus, translated as *The Mirror of Stones,* a description of "the Nature, Generation, Properties, Virtues and various Species of more than 200 different Jewels, precious and rare Stones," along with "certain and infallible Rules to know the Good from the Bad, . . . the Real from Counterfeits," for those who purchase stones for "Curiosity, Use, or Ornament" (see fig. 4).[47] Dedicated to Cesare Borgia, it is a foray into the science of stones. As Leonardus writes in the introduction, "My Purpose therefore, in this little Book, is to treat minutely of stones. For in stones there are many Things to be consider'd with respect to their Essence. As first, the Matter; also, their Virtues, the images impressed on them. Therefore, this Book is entitled *The Mirror of Stones* . . . that as a Mirror, or Looking Glass, truly represents the Images of Things set before it."[48]

Despite this practical and factual intent, *The Mirror of Stones* subscribes to the astral and magical properties of gems. In early sixteenth-century Europe, stones were thought to prevent fire, ward off chills, detect poison, cure melancholy, stop hemorrhaging, improve eyesight,

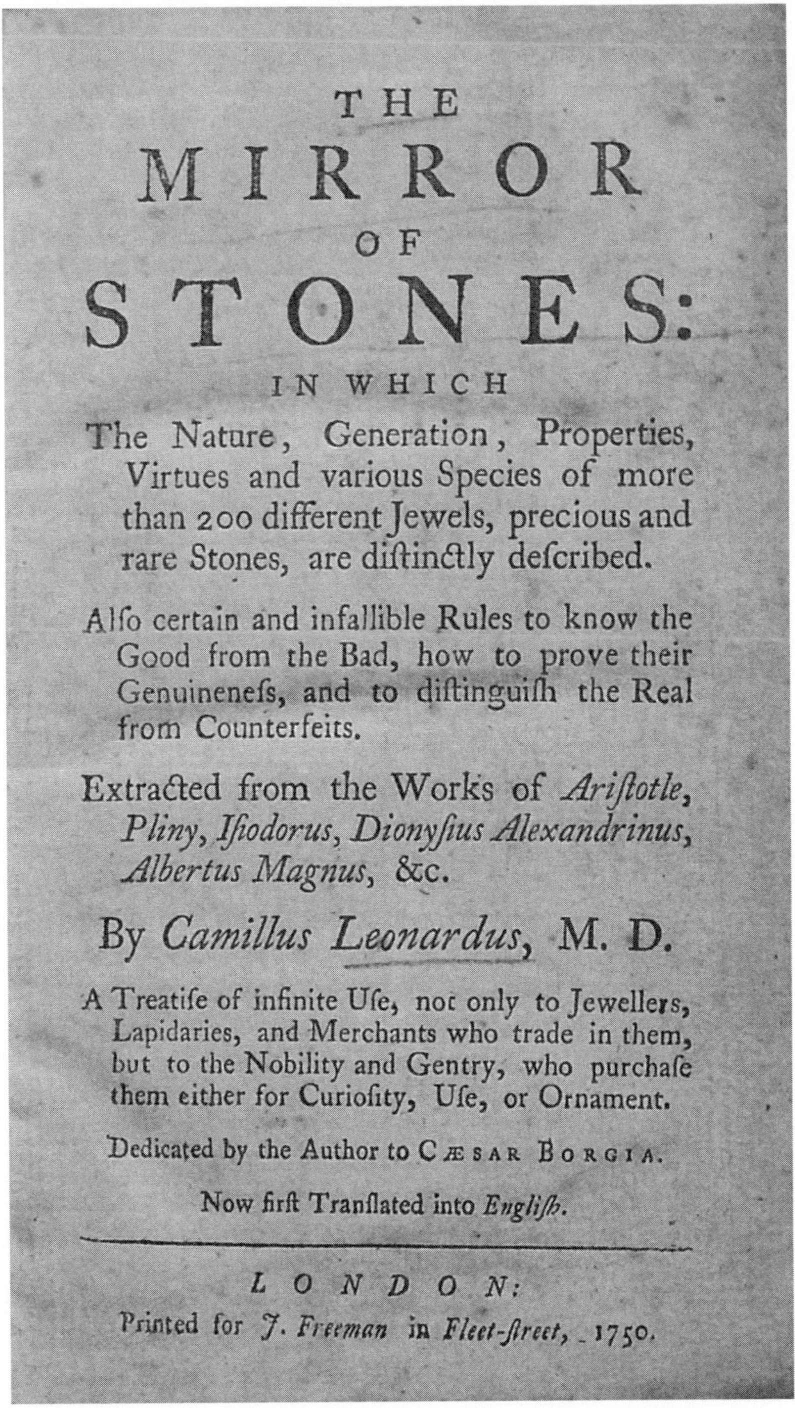

THE
MIRROR
OF
STONES:

IN WHICH

The Nature, Generation, Properties, Virtues and various Species of more than 200 different Jewels, precious and rare Stones, are diftinctly defcribed.

Alfo certain and infallible Rules to know the Good from the Bad, how to prove their Genuinenefs, and to diftinguifh the Real from Counterfeits.

Extracted from the Works of *Ariftotle*, *Pliny*, *Ifiodorus*, *Dionyfius Alexandrinus*, *Albertus Magnus*, &c.

By *Camillus Leonardus*, M. D.

A Treatife of infinite Ufe, not only to Jewellers, Lapidaries, and Merchants who trade in them, but to the Nobility and Gentry, who purchafe them either for Curiofity, Ufe, or Ornament.

Dedicated by the Author to CÆSAR BORGIA.

Now firft Tranflated into *Englifh*.

LONDON:
Printed for *J. Freeman* in *Fleet-ftreet*, 1750.

FIGURE 4. Title page of Camillus Leonardus, *The Mirror of Stones* (London: J. Freeman, 1750).

offer protection, and increase fertility, among other benefits.[49] They were given astronomical significance and were thought to carry rays from the stars, planets, and constellations that could, at auspicious times, affect cures.[50] The emerging geological sciences challenged but did not completely erode these beliefs.[51]

Leonardus's description of turquoise focuses on the stone's physical properties, particularly its "colour . . . very agreeable to the sight," and dismisses the "vulgar Opinion that it is useful to Horsemen, and that so long as the Rider has it with him, his Horse will never tire him, and will preserve him unhurt from any Accident."[52] Still, the third book of Leonardus's work, which the eighteenth-century English translation does not include, covers the magical properties of engraved gems and the planetary associations of ring stones, such as turquoise. Writing on "the peculiar sympathies of precious stones toward the planets," Leonardus identifies turquoise as the gem of Saturn. The association, he suggests, is due to the fact that both are of "cold earth" (*terra frigida*), sharing composition and color. A turquoise stone set on a ring of lead (Saturn's metal) was a zodiacal talisman that could affect one's spirits, even curing the symptoms of melancholy (see fig. 5).[53]

Because stones were thought to possess various therapeutic and curative qualities, remedies were concocted by powdering them. Turquoise was especially prized for this use due to the widespread belief that it was a general antidote to misfortune. The mid-sixteenth century French Catholic physician François de la Rue (also known as Franciscus Rueus) wrote in his lapidary text *De Gemmis* of the "wonderful" and "remarkable" properties of turquoise, concluding that "the stone is gifted with a certain divine power . . . in portending omens," which could "defend from danger the one wearing it, as if it were a certain natural amulet and antidote against bad luck."[54] European naturalists sought to discover the stone's other remedies. In 1563, Garcia de Orta, a Portuguese naturalist and physician in the western Indian port of Goa, identified turquoise as a precious stone that came from Iran and referred to its medicinal uses in his *Colloquies on the Simples and Drugs of India*, a gazetteer on botanical substances in India whose fifty-seven chapters are in the form of colloquies between the author and a Spanish physician named Dr. Ruano. In the chapter on precious stones, Ruano inquires, "Tell me whether the turquoise is used in medicine?" To which Garcia de Orta responds, "Some people tell me that it is, others that it is not. . . . Among the Moors all say that it is used in medicine."[55] De Orta also tells Ruano that "in Arabic *p* and *f* as letters are like

FIGURE 5. The planetary associations of ring stones and metals, according to Camillus Leonardus, *The Mirror of Stones* (London: J. Freeman, 1750), 259. Turquoise and lead belong to Saturn (top left).

brothers. . . . *Ferruzegi* means a turquoise, or of a turquoise, for *Puruza* is a turquoise in Arabic, of which there is a great quantity in Persia."[56] Leaving aside this sketchy knowledge of the Persian and Arabic languages, Ruano replies with gratitude, "Truly for this alone one would wish to come to India, but if I did not find you perhaps I would not say that. From this time forward when I find *Ferruzegi* in Avicenna or in any book of the Arabs, I shall understand it to be a turquoise."[57]

By the seventeenth century, a more critical view of turquoise, its origins, and its trade appeared in European mineralogical texts. In a chapter of *Le Parfaict Ioaillier, ou Histoire des Pierreries* (1608), Boetius de Boot, the imperial physician and alchemist to Holy Roman Emperor Rudolph II (1552–1612), describes the stone's global commerce, specifying that the highest-quality and bluest gems originated from "Persia," in the "Orient":

> There are two kinds of turquoise; Oriental and Occidental. The Oriental is that whose color is composed more of blue than of green. The Occidental is that which is more green or which whitens uncommonly. The first kind occurs in Persia and the eastern part of India. . . . In Persia it grows among black rocks, as if it might be their excrement or transudation; and there it occurs in great quantity. Specimens that I have seen rarely surpass the size of a walnut. . . . The Oriental stones, moreover, are divided into two kinds. One kind retain perpetually their color, and these are called of the old rock. The other kind slowly lose their color and become green; and these are called of the new rock.[58]

Following the "discovery" of the Americas and the conquest of the Aztec Empire, Spanish explorers described turquoises, called *chalchihuitl* in Nahuatl, that they found there. But these "Occidental stones," which were given as tribute in mosaics, were not as valued as the gem-quality ones of Iran and did not enter the global economy as commodities or circulate as widely. Not only were they too green for early modern European consumption, which esteemed turquoise as a sky-blue jewel, but their specimens were already patterned as tesserae in objects of Aztec regalia and more rarely found as stones. On the market, the sky-colored stones of Nishapur were the only specimens true to the name of *turquoise*. Because of its rare color, de Boot claimed, this turquoise was in high demand across Europe:

> The turquoise possesses such authority that no one thinks his hand well adorned, nor has satisfaction in his luxury, unless he wears a fine one. Yet the women are not accustomed to carry this precious stone, because it is not sold at a great price, inasmuch as it is brought in abundance from the Orient.

Its pleasing color gives its value; yet jewelers consider whether its color is going to fade. Those the size of filberts, of a fine sky blue shade and not discolored by black veins, bring 200 thalers, or even more. The smaller ones are lower in value, and their breadth establishes their price. Those the size of a large pea bring 6 thalers. Preferable to all others are those which perfectly express the greenish cast of clear gray and show an agreeable greenish-blue color diluted by the color of milk. Those with black veins, or which are too green or too milky, are of no value.[59]

The shade of turquoise, according to de Boot, ranged from "skye blue" to "air green": "Among the opaque precious stones, the finest of all is the turquoise. . . . It is known among all nations by this name, because it is brought from Turkey hither. . . . This precious stone has a color composed of green, of white, and of blue; and if it is fine, a gray-green color, commonly called air green."[60] The blue green of turquoise had such appeal, de Boot claimed, that the glassmakers of Venice imitated its color and counterfeiters in France fashioned simulated stones from chrysocolla, introducing veins to pass it off as real. Jewelers had also devised means and artifices for restoring the color of faded turquoise, some soaking the stones in water of chrysocolla mixed with the pigment of ultramarine (made of lapis lazuli) and others sanding them with emery and polishing them with tripoli until the color of the unoxidized layers beneath showed through.[61]

Thomas Nichols of Cambridge University described specimens of turquoise from Safavid mines in his manual *Lapidary; Or, The History of Precious Stones,* published in 1652. He confirmed, following de Boot, that there were two varieties—stones of "skie colour" from the "Orient" and those from New Spain, "of an obscure green colour, with an ungratefull aspect."[62] Nichols noted that "the colour of this stone doth set its price. . . . It is of great esteem with Princes and much pleasure they take in its beauty; and it being set in gold they wear it on their fingers." The most prized turquoise "sold for two hundred crowns a piece and more."[63] Gem-quality turquoise came from mines in Safavid Iran: "The Orientall ones are brought from *Persia* and from the *Indies* into *Turky,* and into these parts; these are seldom bigger than a filberd and very rarely seen so big as a walnut. . . . Boetius saith that he never saw one of these gems bigger than a filberd. . . . Some of the Orientall ones are said to keep their colour perpetually, and those are called *Turkies* of the old rock, and some of these gems are said by degrees to loose their colour and grow greenish, and these are called *Turkies* of the new rock."[64] While the Persian name given for the stone was *firuza,* in Europe it was called *turquoise,* because it was "brought from the Turks."[65]

The stone's rare and sometimes vanishing "skie-coloured" shade was held in such high esteem that the Venetians, Nichols reported, had developed ways of imitating and recovering it, the latter by rubbing with "oyl of Vitrioll."[66] But turquoise needed no trickery to be appreciated, as he noted:

> The *Turkey Stone* is a very hard gemm of no transparency, yet full of beauty, as giving the grace of its colour in a skie colour out of a green, in which may be imagined a little milkish perfusion; Indico will give the perfect colour of it, and Verdigrease hath a perfect resemblance of it; and a clear skie colour free from all clouds will most excellently discover the beauty of a *Turkey Stone.* Non-transparent stones, and wholly shadowed gems admit of no foyls, therefore nothing concerning them must be expected. The *Turkey Stone* is throughout of the same beauty, as well internally as externally; it wants no help of tincture to set it off in grace, the constancy of its own beauty without any extraneall help is the support of it and beareth it up against all defects. It is an excellent gemm of a most simple substance, in every part like it self, most pure in colour, and without spot, and the constancy of its beauty is a sufficient commendation for it self.[67]

Nichols described the stone as "skie coloured" in the daytime, which he considered to be among the "originall" colors.[68] Mixtures of "white, black, blue or skie colour, yellow, red or vermillion or fiery red" created almost all other colors. Natural substances such as precious stones and plants, he noted, brought into the world "the originall of the varietie of beauty in colour." Color was contained "in the substantiall matter of all things," and their various colors distinguished the natural things on earth.[69]

By the seventeenth century, descriptive mineralogical texts such as those of de Boot and Nichols, while still relating many traditions regarding the mysterious properties of turquoise and other precious stones, were casting these older beliefs into doubt.[70] As stones became global commodities, they were demystified and detached from older meanings, valued for their sheer physical and material properties and their monetary worth as goods. De Boot wore a fine blue engraved turquoise ring constantly for many years and was willing to entertain the idea that it protected him from harm, but he ultimately resisted attributing special powers to precious stones: "I am convinced that naturally this precious stone cannot prevent the accident being harmful, nor attract upon itself the evil. It is necessary then to attribute these results to an occult agent. . . . At least I can certainly assert that I have never believed, nor do I now believe, that such a thing ever naturally occurs to the turquoise."[71]

Changes in the stone's color, de Boot further suggested, had less to do with its supposed metaphysical properties and more with "natural fact," primarily its exposure to different oils and substances during use.[72] In his discussion of "the nature, faculties, and properties of the Turchoys Stone," Nichols also cast doubt on the "many strange things beyond faith . . . reported concerning the virtues of this stone, which nothing but excess of faith can believe." Although he found turquoise to be "ceruleous like a serene heaven . . . delightfull to the eye," he doubted that it would preserve its owner from falls or diagnose his or her health.[73]

Into the early nineteenth century, turquoise was still an object shrouded in uncertainty and mystery among European jewelers and consumers. A line of European authors had argued for some years that it was made of the petrified bones and teeth of animals, referring to it as *odontolite,* and could be found in places such as Languedoc in France, while others searched for methods to produce counterfeits of the opaque blue stone, which only increased confusion in European markets as to its true nature. In 1715, the Parisian naturalist and savant René Antoine de Réaumur identified turquoise as either an "Oriental" stone from Persia or an "occidental" piece of petrified bone or teeth found in parts of Europe.[74] Writing in 1790, the French chemist Jean-Antoine Chaptal claimed in his *Éléments de Chimie* that "turquoise stones are merely bones coloured by the oxides of copper."[75] In an 1806 volume of *Annales de Chimie,* the French chemist and pharmacist Edme-Jean Baptiste Bouillon-Lagrange seems unsure whether to classify turquoise as a stone or the petrified parts of animals.[76]

European scientific classifications of precious stones such as turquoise now departed from the descriptive style of the past and came to unlock the physical composition and chemical properties of mineral substances. This shift had been under way since the sixteenth century, as seen in *De Re Metallica* (1556) by George Bauer, known by the pen name Agricola, a text that systematized the modern science of mineralogy and mining as a classification of the chemical and physical composition of the substances of the earth.[77] Rocks, crystals, and minerals were increasingly grouped by their external characteristics, such as hardness, composition, chemistry, and color.[78]

In time, however, turquoise was firmly established in natural histories—and markets—as one of the exotic mineral substances of the earth found in the faraway mines of the East. The acclaimed English naturalist and artist James Sowerby identified it as a stone of pale blue found nowhere but in "Persia" in his *Exotic Mineralogy,* an illustrated catalogue of minerals whose two volumes were printed in London in 1811 and 1817: "True turquoise . . . occurs in veins and tuberose masses in a grayish black stone. There are probably several mines that are worked for it, all situated in Korosan, in the northwest part of Persia. . . . The genuine Oriental Turquoise or Birjusa, is found in no other country." He also illustrated a specimen "from a mine at Nichassur, in Korosan," in a color plate in the first volume (see plate 9).[79]

Nineteenth-century geologists and other earth scientists turned the microscope on turquoise, finding it to possess a soft and porous texture with a waxlike luster and a high index of light refraction. Its hardness was measured at about 6 on the Mohs scale, soft in comparison with other gems, and its specific gravity between 2.60 and 2.88. Its surface could be scratched by quartz and polished by sand. The stone has an opaque sky-blue color. When heated by a blowpipe, it takes on a glassy brown appearance and colors the flame blue green.[80]

Color itself came to be explained as a physical phenomenon—the perception of light's particular interaction with each object on which it falls. In 1666, Isaac Newton revealed in an experiment that white light contains all the colors of the rainbow, which bend at fixed angles when passed through a prism—or drops of rain. As he noted in his book *Opticks: Or, a Treatise of the Reflexions, Refractions, Inflexions and Colours of Light,* this finding indicates that color is the physical perception by the human eye of light refracted in the external universe by every object with which it comes in contact.[81] Newton's classification of the color spectrum in white light—red, orange, yellow, green, blue, indigo, violet—also facilitated the description of the hues of rocks and stones.[82] To be sure, those who made the case that colors are not real and would not exist if not for the imagination challenged this detached view of them. Johann Wolfgang von Goethe's *Theory of Color,* published in 1810, claims that colors are an anthropological and cultural phenomenon.[83] In scientific and mineralogical literature on precious stones, however, the conception of color as a physical characteristic became pervasive. In *Traité de Minéralogie,* René Just Haüy, a professor at the Museum of Natural History in Paris, developed the theory that all crystals have an orderly internal structure of mineral composition that determines their external shape and their brilliant reflections.[84]

As for turquoise and its color, the sky blue of the stone came to be defined by its chemical composition—the opaque celestial blue, it was known, derived from an oxide of copper.[85]

FADING COLORS

Turquoise was among the colorful commodities of Asian origin that spread the material culture of blue and gave substance to the construction of the color in the early modern world. Extracted from the mountains of eastern Iran, the stone was exported along the crossroads of the Islamic empires of Eurasia—founded by pastoralists who had moved from the steppe to the sown—where it represented imperial conquest and was replicated as the central color in the architecture of metropolitan monuments and urban spaces, to Europe, where it came to be known as a rare and exotic Persian stone and a distinct color of blue.

The early modern turquoise trade flowed through contacts and exchanges between two imperial worlds—the Islamic tributary empires of Eurasia, and European commercial empires, including East India companies. In the tributary economies of Muslim empires, turquoise was ingrained in the assertion and display of imperial power, rooted in the synthesis of Central Asian and Persianate cultures, in places where sovereignty was layered and mediated among semiautonomous subjects and populations. The establishment by European commercial empires of colonies in Asia and the New World—providing easier access to resources—opened the way for the trade of an array of exotic goods, including precious stones and earths such as turquoise. But removed from their context in the eastern Islamic world and placed in the markets of Europe, turquoise and its trade did not carry the same import, and along the way the stone lost its meaning as the regalia of imperial power and a celestial, sacred earth. In the markets of Europe, turquoise was just one of a number of global commodities to be consumed, classified, and collected, and stripped of its mystique, it was surpassed in value by silks, spices, and metals such as silver and gold, which became the basis of the modern world economy.

Even the color of the stone, once seen as the essence of its wondrous virtues and properties, was demystified and came to be understood strictly according to its physical and chemical composition. With the fall of the Safavid dynasty in 1722 and the subsequent decay of the mines of old rock outside Nishapur, the culture of turquoise as an object of interimperial exchange and power in Islamic Eurasia also began to wane. In the nineteenth century, prospectors attempted to reclaim and revive the Eurasian turquoise trade and its culture but found that the economy of the stone—its routes, values, and meanings across the world—had lastingly changed.

CHAPTER 5

The Other Side of the World

The nineteenth-century gold rush and global quest for precious metals and stones transformed the Eurasian turquoise trade. Imperial projects to order and reclaim nature and its resources attempted to revive the lost turquoise mines of the Near East. At the start of the century, the ancient turquoise deposits of the Egyptian Sinai were largely forgotten, known and visited only by local Bedouins. But European explorers surveying the ruins, monuments, and environments of ancient Egypt soon "rediscovered" these mines, and in the 1840s a British cavalry officer, Major Charles Macdonald, set out to reopen them. In 1882, on a tour of the eastern Iranian province of Khurasan, Nasir al-Din Shah Qajar determined to reestablish state regulation over the once-lucrative turquoise mines near Nishapur. He appointed a superintendent of the mines and sponsored the production of scientific texts on the geology and natural history of stones. Through such imperial mining projects and the production of lithographed and printed works of geology and natural history, the Qajars strove to reclaim Iran's subterranean mineral resources, and to meet the demands of the world economy.[1] However, Mesoamerican turquoise deposits in the deserts of the American Southwest were again found and opened up, becoming accessible sources of turquoise from the New World to rival the sky-blue Persian stones on the global market.

MINERALS AND MARKETS

The nineteenth century—powered by telegraphs, modern roads and rails, steam, and print—saw the emergence of a more thoroughly globalized economy.[2] The increased supply of and demand for gems, as well as more rapid means of transporting them that linked together geographical spaces as never before, brought precious stones into mass consumption. The economy of this nineteenth-century trade penetrated more deeply as stones came into the possession of different strata of society. Responding to the global demand for sartorial and other decorative jewels and ornaments, workers around the world mined the earth or ventured to the depths of the sea in search of precious stones and gems.

On entering this modern world economy of gems, turquoise came into prevalent demand among European consumers as a precious stone from the Orient, a rare object from a place of romance and fascination. By the early nineteenth century, it was in ample circulation among consumers in Europe, where it was "in fashion, the demand for the stone considerably greater than the supply."[3] It was widely known that the turquoise of commerce came from the mines of Nishapur in eastern Persia. In an 1813 treatise on the natural and commercial history of precious stones, the British jewel merchant John Mawe specifies that "the most valued kinds of Turquois come from Persia," where "the Grandees of Persia, and of the adjacent Mahometan states" had long favored it.[4] The demand for the stone, according to Mawe, was due to its extraordinary color, the "essential character of Gems," at which he marveled: "Opake, and not admitting of a very high polish, there is nothing but the agreeable tone of its colour which . . . it possesses in no inconsiderable degree, especially by candle-light."[5] Other scientists and authors confirmed that the "true" turquoise, of gemstone quality—the stone that had become an Oriental novelty and rarity to be collected in nineteenth-century Europe—originated in Iran and was greatly esteemed by Asians, who used it as ornaments, talismans, and jewelry.[6]

Turquoise turned up in the drawers of cabinets of curiosities and in displays of collectors' precious stones—a rare Persian jewel from the bazaars of the East. The turquoise ring became a widely coveted exotic in Europe, a curio and object of art from far away to be collected and displayed. The stone placed the foreign mysteries of the East in European consumers' hands. In the process, it retained aspects of its astral significance, being associated with prosperity and good fortune, particularly for, it was thought, those born in December, who were advised, "Place on

PLATE 1. A Safavid parade helmet (*kulakhud*), possibly looted from Tabriz by the victorious Ottoman armies after the Battle of Chaldiran. Topkapi Sarayi, Istanbul.

PLATE 2. *(opposite)* The Mughal emperor Jahangir, atop the world with a turquoise ring on his right hand, embracing the Persian shah 'Abbas I, by Abu'l Hasan. Mughal India, c. 1618. Freer Gallery of Art, Smithsonian Institution.

PLATE 3. The Pattern of the World—Naqsh-i Jahan: the *maydan* of the turquoise city of Isfahan, consisting of (from left to right) the Shaykh Lutfallah Mosque, the Masjid-i Shah, and the 'Ali Qapu Palace. Pascal Coste, *Monuments Modernes de la Perse Mesurés, Dessinés et Décrits* (Paris: A. Morel, 1867).

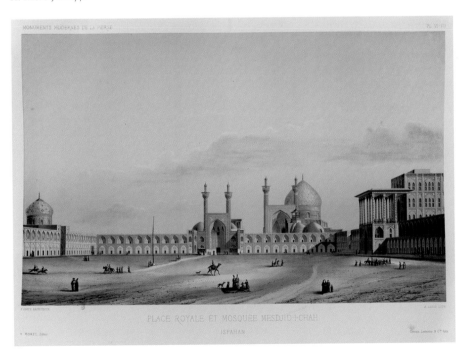

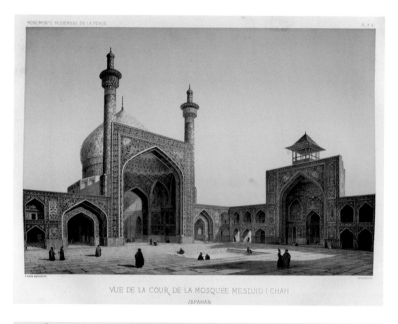

VUE DE LA COUR DE LA MOSQUÉE MESDJID-I-CHAH
JSPAHAN

MOSQUEE MESDJID - I - CHAH.

PLATE 4. The inner courtyard of Isfahan's Royal Mosque—Masjid-i Shah. Pascal Coste, *Monuments Modernes de la Perse: Mesurés, Dessinés et Décrits* (Paris: A. Morel, 1867).

PLATE 5. The seven colors (*haft rang*) of heaven: detail of the ceramic tiles of the Royal Mosque—Masjid-i Shah. Pascal Coste, *Monuments Modernes de la Perse: Mesurés, Dessinés et Décrits* (Paris: A. Morel, 1867).

F. COSTE, ARCHITECTE   AD. LEVIE LITH.

MOSQUÉE DU MEDRECEH MADERI-CHAH-SULTAN HUSSEIN

MINARET ET DÉTAILS

JSPAHAN

PLATE 6. Detail of a turquoise minaret of Madrassa-yi Madar-i Shah. Pascal Coste, *Monuments Modernes de la Perse: Mesurés, Dessinés et Décrits* (Paris: A. Morel, 1867).

Sindh & Tombs of the Talpur family, Ameers of Sindh, at the period of the conquest by Sir C. Napier 1843

PLATE 7. *(opposite, top)* Blue city of Sindh: interior of the Shah Jahan Mosque in Thatta. T. Wingate, Queen's Royal Regiment, London, Asia, Pacific, and Africa Collections WD 1033 (1839), India Office, British Library.

PLATE 8. *(opposite, bottom)* Indian variations on a theme: tombs of the amirs of Hyderabad, Sindh. Henry Francis Ainslie, South and Southeast Asian Collection IS.27–1963 (1852), Victoria and Albert Museum, London.

PLATE 9. Genuine turquoise: a color sketch of the Persian stone in the raw. James Sowerby, *Exotic Mineralogy: Or, Coloured Figures of Foreign Minerals as a Supplement to British Mineralogy,* vol. 1 (London: Benjamin Meredith, 1811), table 93, History and Special Collections for the Sciences, UCLA Library Special Collections.

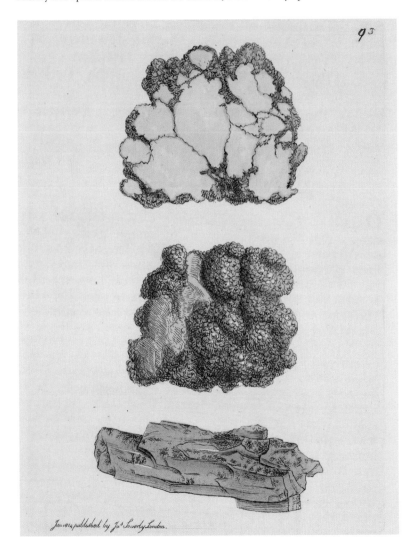

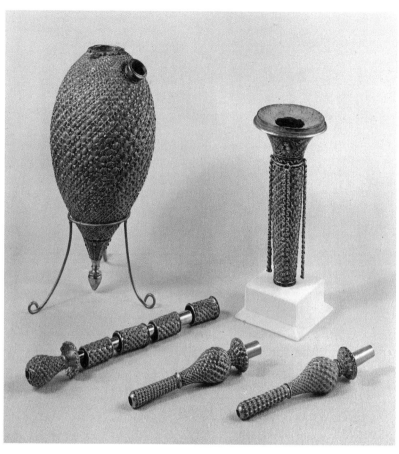

PLATE 10. The shah's *qalyan*: a turquoise and ruby water pipe that belonged to Nasir al-Din Shah Qajar. Treasury of National Jewels and Precious Stones, Central Bank of Iran, Tehran.

your hand a turquoise blue / Success will bless whatever you do."[7] The trade of the stone extended beyond merchants, mineralogists, and royals "to include the general public," because unlike the more costly ruby, emerald, and diamond, it was "sufficiently abundant to come within the reach of the average buyer."[8] As the London-based lapidary Harry Emanuel of Bond Street observed, "The Persian turquoise is much used in jewelry, and a great number are sold here."[9] By the mid-nineteenth century, a ring stone realized between ten and forty pounds in London and Paris, while larger stones of good depth could be valued at four hundred pounds. Smaller stones cost between six dimes and twenty shillings.[10]

As turquoise spread throughout European cities and markets, it also turned up in the pages of Orientalist texts and other works. Translations of extracts from Muhammad ibn Mansur's *Javahirnama* and other books of gems appeared in European journals, including an 1820 edition of the *Annals of Philosophy* which reminded readers that "the knowledge of precious stones first arrived with them from the East . . . where the mines were situated" and echoed Ibn Mansur's praise for the turquoise of Nishapur.[11] Lapidary texts acquainted readers and consumers with the culture of turquoise as a sacred and regal stone among Islamic empires. Emanuel reported that "the monarchs of the East, with their fondness for display and pomp . . . decorated their horse trappings, their thrones, and their persons with gems, long before they knew how to cut them; and they attributed, as they even now attribute, magic and talismanic properties to them." He further observed that "turquoise is much used in oriental countries for ornamenting harnesses, girdles, swords, daggers, and pipes, also for amulets and charms. It is also frequently engraved with the name of Allah, a verse of the Koran, or some device, and then filled in with gold. . . . The Shah of Persia is supposed to have in his possession all the finest gems, allowing those only of inferior quality to leave the country."[12]

Turquoise also featured in works of Orientalist fiction. Charles Hetherington's *Selim, the Nasakchi*, for instance, touts the beautiful stone of the tumultuous and exotic East:

And soon 'tis known, rebellion's meteor shines
In that wide province, rich in turquoise mines,
Khorasan—Persia's eastern land, which lies
Bord'ring the realms 'neath India's sultry skies.[13]

The stone's color itself became a commodity on the surfaces of printed and lithographed books and art. A phosphate pigmented by aluminum and copper, turquoise came to specify a particular shade of blue—that of the sky—which joined the chromatic spectrum of the printing press.

FRAGMENTS OF HATHOR

The earliest attempts to reopen turquoise mines to meet the nineteenth-century demand for the stone occurred in Egypt. The French expedition there from 1798 to 1801 marked the beginnings of the project to unearth, recover, and revive the Egyptian past. Scientists and savants in the expedition uncovered the ruins of Egyptian antiquity, which they described in the twenty-four-volume *Description de l'Egypt*.[14] But the expedition excavated more than monuments and hieroglyphics. In addition to archaeological and cultural projects, it generated environmental narratives about the land and its natural resources. The *Description* presents a natural history of Egypt and the Nile, revealing their flora, fauna, and mineral strata in zoological, botanical, and geological detail. In the course of their work, French scientists espoused a belief that the land of the Nile could be reclaimed, developed, and restored to its monumental past.

Such ideas persisted long after the French had departed. The Ottoman provincial governor Muhammad 'Ali Pasha (r. 1805–48) resumed the developmental project of centralizing his state and extending its control over Ottoman Egypt's environments. The most well known of Muhammad 'Ali Pasha's environmental efforts was the control of the Nile floods through perennial irrigation, begun in the 1820s, in order to produce the cash crop of cotton for the world market.[15] The nineteenth-century project of converting nature worked in tandem with the emergence of a fully globalized world economy, culminating in the construction of the Suez Canal in 1869, considered the greatest feat of engineering the world had seen, which linked the Mediterranean and the Indian Ocean (via the Red Sea) and cut the distance between Bombay and London in half, expanding the world economy and more closely connecting the British imperial world from India to Egypt. The convergence of empire and environment in nineteenth-century Egypt also proved a catalyst for the rediscovery of the ancient turquoise mines of the Sinai.[16]

Early interest in Egyptology and researches into Egypt's mineral riches were entwined in the European quest to identify the true route of Moses and the Israelites thorough the Sinai during the Exodus.[17] Along the way, European travelers, Orientalists, and explorers rediscovered the ruined mines of turquoise, known as *mafkat*, at the ancient sites of

Wadi Maghara and Sarabit al-Khadim, in south-central Sinai. These mines in the Sinai Peninsula were likely first unearthed by Bedouins. Subsequently, the pharaohs sent expeditions there, among the local Bedouin populations, to secure mineral and other natural resources necessary for their monumental building projects in the Nile delta.[18] Outside the mines, inscriptions, stelae, and graffiti record the details of mining expeditions and attest to dynastic exploits in mineral extraction.[19] The turquoise mines in Wadi Maghara, or the Valley of Caves (or Grottoes), were the first to be exploited, as the inscriptions found in the valley record. Captives worked them for more than two thousand years.[20]

By around 2100 B.C.E., the veins of turquoise in Wadi Maghara were exhausted and its mining colonies vanished.[21] With the demand for the stone still great, explorers soon discovered untouched turquoise deposits about ten miles away, at Sarabit al-Khadim, which successive expeditions of the pharaohs exploited. The workers dedicated the mines there to Hathor, the goddess of the turquoise land, and built a temple in her honor, where prayers, vows, and offerings could be made.[22] The last pharaoh to leave an inscription at Sarabit al-Khadim was Ramses VI (r. c. 1145–1137 B.C.E.). With the subsequent depletion of the turquoise stratum, the Temple of the Turquoise Goddess was abandoned, until Egyptologists and colonial adventurers rediscovered it in the nineteenth century.[23]

At that time, the offerings that ancient Egyptians had made to Hathor were in fragments strewn around the shrine. They included beads and now-broken pottery dedicated to the Turquoise Goddess by those who had sought her favor in the prospect of mining.[24] Explorers and archaeologists also visited and described the remains of the nearby mines. In 1828, a local guide accompanying the French antiquarian Léon de Laborde discovered turquoise stones while searching amid the ruins at Sarabit al-Khadim.[25] Nearby residents regarded the value of turquoise as purely medicinal or religious, as when melted with other minerals and metals and offered to their gods.[26] Noticing the abundance of turquoise stones washed down from the hills and brought to the surface by rains, Laborde speculated that the area was rich in mining prospects, ripe for the production of stones for the world market. On the possibility of reestablishing an Egyptian turquoise industry, he wrote, "A person spending a few days on this mountain, where he will be exposed to no danger, and which is not more than six days' journey from Cairo, might make a large collection of turquoises, which, though not to be ranked among the best of precious stones, nevertheless possess a certain value."[27]

The geology and the red sandstone landscape of the Sinai also struck nineteenth-century travelers. In the illustrated four-volume survey *Picturesque Palestine*, Charles Wilson, a former engineer for the Palestine Exploration Society, romantically describes the region, including its mineralogical wealth: "Very fantastic are the shapes and gorgeous is the colouring of the mountains in this district; the valleys are narrow and steep-sided, whilst there are many undulating barren plains of gravelly sand found at their mouths. In this formation are veins of iron, copper, and turquoise, in the mines of which Egyptian captives from the far south or from the northern Hittite country pined away their life, generation after generation" (see fig. 6).[28] At the time, Sarabit al-Khadim was "an extensive plateau broken up by deep ravines and rising knolls . . . a heap of ruins—hewn sandstone walls, with broken columns and numerous stelae."[29] Wilson noted that the remains of the turquoise mines there were still to be found in "great heaps of slag and other vestiges."[30]

Inscriptions at Wadi Maghara and Sarabit al-Khadim had left a record of Egyptian expeditions in the quest for turquoise. Engraved tablets in the mountainsides memorialized these mining campaigns, recording the names of the engineers and detailing the work of the miners in the Sinai.[31] A stela at Sarabit al-Khadim commemorates a superintendent of the mines who arrived at his post one winter in the reign of the Twelfth Dynasty (2466–2266 B.C.E.). The inscription claims that he never once left the mines while on the job and calls on others to join him:

> If your faces fail the goddess Hathor will give you her arms to aid you in the work. Behold me, how I tarried there after I had left Egypt; my face sweated, my blood grew hot, I ordered the workmen working daily and said unto them there is still turquoise in the mine and the vein will be found in time. And it was so; the vein was found at last and the mine yielded well. When I came to this land aided by the king's genii I began to labour strenuously. . . . I toiled cheerfully; I brought abundance, yea, abundance of turquoise, and obtained yet more by my search. I did not miss a single vein.[32]

The researches of the Lepsius archaeological expedition in 1828–29, commissioned by Prussia to explore Egypt and the Sudan on the recommendation of Alexander von Humboldt, had revealed the purpose of these monuments and their connection to the local mines.[33] By deciphering their hieroglyphs and other images, Egyptologists pieced together the work of the miners in the turquoise land. For instance, a stone tablet at Wadi Maghara portrays a group of miners overseen by a guard armed with a bow and arrow, demonstrating that gangs of captives had

FIGURE 6. Rediscovering ancient Egypt: the turquoise mines of Wadi Maghara, the Valley of Grottoes. Charles William Wilson, *Picturesque Palestine: Sinai and Egypt,* vol. 4 (London: J.S. Virtue, 1884), 59.

worked the mines.[34] Hieroglyphics etched on the rocks above the mines show the tools that the miners had used—rough hammers and chisels.[35] Nineteenth-century travelers at Wadi Maghara also found traces of Egyptian excursions sent to extract iron, copper oxides, and turquoise, to be exported to the delta. These "expeditions started from different points of the valley, swept down upon the peninsula, and established themselves by force in the midst of the districts where the mines lay . . . in the region of sandstone."[36] The miners had searched for veins of turquoise in narrow connecting tunnels, galleries, and halls below the ground, using rough flint and stone picks and chisels, and other tools to separate it from the surrounding rock.[37] According to Gaston Maspero,

> The workers used stone tools . . . to cut the yellow sandstone. The tunnels running straight into the mountain were low and wide, and were supported at intervals by pillars of sandstone left *in situ*. These tunnels led into chambers of various sizes, whence they followed the lead of the veins of precious mineral. The turquoise sparkled on every side—on the ceiling and on the walls—and the miners, profiting by the slightest fissures, cut round it, and then with forcible blows detached the blocks, and reduced them to small fragments, which they crushed, and carefully sifted so as not to lose a particle of the gem.[38]

A "few hundred" men had worked the mines at Wadi Maghara. Seasonal expeditions would set off following the flooding of the Nile in late summer and early fall, to mine for turquoise in the cooler winter months. Before the hot summer began, the turquoise hunters would return to Egypt.[39] "Royal inspectors arrived from time to time," Maspero wrote, "to examine into their condition, to rekindle their zeal, and to collect the product of their toil."[40] When the pharaoh had need of greater quantities of turquoise and other minerals, two to three thousand additional workers poured into the Sinai, and settlements were built to shelter them. One inscription details 734 men in a single mining expedition.[41]

In the mid-nineteenth century, Major Charles Macdonald, a Scottish soldier, archaeologist, and explorer, set out to reopen the Sinai turquoise mines. He left few records, and the sparse traces of his life and work that remain may be found in Heinrich Brugsch's 1866 volume on turquoise mining in the Sinai, *Wanderung nach den Turkis-Minen und der Sinai-Halbinsel*, and among the objects in the collection of Egyptian antiquities in the British Museum.[42] Macdonald was born in the Hebrides Archipelago, off the west coast of Scotland, and later served as a major in the British cavalry. He went to Egypt with his wife in the 1840s and first visited the Sinai in 1845. Seeking the land of the turquoise mines and the Temple of Hathor, he traveled by camel with Bedouin guides to Sarabit al-Khadim. Macdonald remained there for an extended period, excavating more than four hundred objects and making squeezes of inscriptions. By 1847, these objects had reached the Department of Egyptian Antiquities in the British Museum.[43]

His serious interest in reopening the turquoise mines of the ancient Egyptians, as he wrote in one surviving account, seems to have been aroused in the Sinai:

> In the year 1849, during my travels in Arabia in search of antiquities, I was led to examine a very lofty range of mountains composed of iron sandstone, many days journey in the desert, and whilst descending a mountain of about 6000 feet high by a deep and precipitate gorge, which in the winter time served to carry off the water, I found a bed of gravel, where I perceived a great many small blue objects mixed with the other stones; on collecting them, I found they were turquoises of the finest colour and quality. On continuing my researches through the entire range of mountains I discovered many valuable deposits of the same stones, some quite pure, like pebbles, and others in the matrix. . . . The action of the weather gradually loosens them from the rock, and they are rolled into the ravines, and, in the winter season, mixed up by the torrents with beds of gravel, where they are found. Another formation is, where they appear in veins, and sometimes of such a size as to be of immense value. They also occur in soft yellow sandstone, enclosed in the centre, and of surpassing brilliancy of colour.[44]

Macdonald's collection of two hundred specimens of turquoise from Egypt, cut and polished, was put on display at the London Exhibition of 1851.[45] By 1854, he had returned to Wadi Maghara with his wife and perhaps also his son to work the turquoise mines. He hired Bedouins to build a house there from local stone and palm trees, where he kept many tame desert animals. Macdonald lived in desert solitude amid the steep sandstone walls that surrounded the narrow valley of Wadi Maghara. By necessity, he forged close ties with the local population. He spoke Arabic, among other languages, and welcomed Bedouins to his home. Some of them settled down nearby, joining him in building their own dwellings of stone.[46] Macdonald also employed more than three hundred Bedouins in his romantic venture to reopen the abandoned turquoise mines of the ancient Egyptians.[47]

In the Victorian Age, turquoise was in fashion in Europe, and Macdonald's venture was an effort to profit from this demand. But his travails to reopen the mines were ill fated. He worked with his men under difficult and isolated conditions in the red sandstone cliffs of the Sinai amid ruins and hieroglyphics and in temperatures that often reached as high as 110 degrees Fahrenheit.[48] Supplies and information had to come from Suez, four days distant.[49] Macdonald's crews constructed solitary roads and desert tracks to allow for transport and communications to and from the mines.[50] Within a short time he was able to send blue turquoise specimens from Egypt to London for an exhibition of the world's minerals. But they faded to green and white. One stone his men had unearthed was "as large as a pigeon's egg" and purportedly "would have been among turquoises what the Koh-i-noor is among diamonds" had it not lost its color.[51]

By 1857, it seems, Macdonald needed capital to continue mining and was forced to return to London to auction his eclectic collection of antiquities, sculptures, and gems through Sotheby's.[52] In the end, his venture to reopen the turquoise mines proved a failure, as Egyptian stones were known to fade and differed in shade from the esteemed Persian turquoise in great demand among European jewelers. The blue stones of old rock from Nishapur set the standard across the world, and the Sinai stones could not compare. According to one nineteenth-century lapidary text, Macdonald's turquoise changed its hue "in the most rapid and mysterious manner," a metamorphosis from a "fine blue" to "a sickly green or whitish tint." The author continued, "It would be well never to give a large price for any turquoise from this mine," which could be "readily distinguished from the real turquoise de vieille roche by the stratum (in most cases apparent on the back) being of a pale yellowish-red colour, instead of dark brown."[53]

According to Edward Henry Palmer, Macdonald was "tempted by the exaggerated accounts of the wealth of the turquoise mines" and searched for precious stones that "turned out to be of little or no commercial value," leaving him "a ruined and disappointed man."[54] Macdonald continued his enterprise in Wadi Maghara until 1866, when he gave up, left his rough stone house, and moved to Sarabit al-Khadim, where he continued his search for salable turquoise. After one unsuccessful year there, he departed for Cairo, where he lived "on the brink of poverty" until his death in 1870.[55]

Subsequent travelers to Wadi Maghara encountered Macdonald's former workers and the material remains of his life and work at the turquoise mines.[56] The Bedouin there and at Sarabit al-Khadim recollected Major Macdonald and his ill-fated venture. He was remembered as a mining explorer with knowledge and appreciation of the archaeological heritage of the Sinai, "always diligent to take paper squeezes of the inscriptions, to search out fresh ones, and to carefully preserve all that he could find."[57] The stone huts that Macdonald and his miners had occupied still covered the plateau.[58] Palmer came across the sad remains of Macdonald's dwelling and other relics of the life he had led in the Sinai:

He had selected a sheltered spot in the immediate neighborhood of the mines, and, having leveled the ground and removed all obstructing rocks by blasting or cutting them away, had built himself a rude but commodious house. . . . Near the spot we found the remains of "Blackie," the Major's cat—some hair and teeth were all that remained of the poor desert puss. When inhabited and kept in order, the house must have looked pretty and picturesque, but it is now in a very deserted and dilapidated state, and is used as a storehouse by the Arabs.[59]

Traces of the local turquoise economy persisted as well. Wilson recalled entering Wadi Maghara only to be accosted by Bedouins who took him for a merchant seeking the stone.[60] Turquoise mining in the Sinai continued well after Macdonald had departed, often at the expense of ancient monuments that the blasting of rocks destroyed.[61] In the late nineteenth century, during the British occupation of Egypt, European mining companies were formed and set out to develop these mines. They used gunpowder without care for the sculptures, tablets, and inscriptions that commemorated the mines, breaking them up in their own mining for turquoise.[62] While Macdonald had been an archaeologist-turned-miner, with an interest in the preservation of ancient Egyptian monuments, subsequent prospectors were more concerned with turning a profit than with the antiquities and remnants of the past. They destroyed the mines themselves in the quest to extract mass quantities of turquoise. According to William Matthew Flinders Petrie, writing at the turn of the century, "All the faces of the rock which had born the royal inscriptions at the sides of the excavations had been blasted away; and the old galleries had been so widened and destroyed that the roof had at last fallen in immense blocks, and all the rock front over the mines was a fresh surface of broken stone."[63]

REDISCOVERING TURQUOISE IN EARLY QAJAR IRAN

Early nineteenth-century European travelers and Orientalists also made their way to the turquoise mines in eastern Iran, where they chronicled the decline of the stone's trade since the times of the Safavid Empire. The Afshar dynasty (1736–96), although centered in eastern Iran, had a brief interregnum and never got around to unearthing the empire's mineral deposits, recovering its mines, or reclaiming the turquoise throne, and Iran lost its position as a pivot in the global gem trade. The founder of the dynasty, Nadir Shah Afshar (d. 1747), instead devoted his energies to raiding the treasuries of neighboring empires for precious stones and other regalia, boldly displayed in his 1739 siege of Delhi and plunder of the Mughal jewels, including the Peacock Throne and the coveted Kuh-i Nur, or "mountain of light," diamond. By the time the Qajar dynasty ascended to the throne, the most it could accomplish with the turquoise mines of Nishapur was to farm them out to the highest bidder, usually the governor of Khurasan Province, who in turn leased them to local villagers. Over time, imperial neglect, fleeting operation and administration, and widespread efforts at maximizing output—attended by the increased use of gunpowder over the pick—led to a chaotic spree that destroyed the country's oldest and most valuable mines.

In *Travels in Various Countries of the East,* William Ouseley, a British Orientalist and member of his brother Gore's mission to the court of Fath ʿAli Shah Qajar (r.1797–1834) in 1811, notes that he found turquoise in bazaars in "every large town in Persia," as far away from Nishapur as the port city of Bushire.[64] Due to the culture surrounding the stone, including the belief that it was auspicious, it was in widespread use: "The Turquoise is an universal favorite; called *firuzeh,* or more properly *piruzeh,* by the Persians, who believe that to look on it when first awake in the morning, ensures prosperity, and highly strengthens and preserves the sight during the whole day. Its efficacy, however, in this respect, does not altogether depend on magnitude; and to the lower classes a *firuzeh* not so large as a grain of wheat (but seldom perfect) is sold for one shilling. Such rings are daily seen on the coarse fingers of muleteers, grooms, and tent-pitchers."[65] On a later mission to the Qajar court, that of the Comte de Sercey in 1839, the French Orientalist artists Pascal Coste and Eugène Flandin surveyed and sketched the sky-blue cities of Iran, revealing the architecture and the material culture of turquoise.[66]

In the winter of 1822, the intrepid Scottish traveler and writer James Baillie Fraser visited and reported on the turquoise mines outside Nishapur while serving on a diplomatic mission in Qajar Iran. He "designed to make a venture with turquoises" for the Bukhara market.[67] Approaching the two mining villages of the Maʿdan, which lay about forty miles northwest of Nishapur, Fraser saw a rocky land of "coloured earths," soils of gray, yellow, ocher, red, and brown.[68] A "porphyritic and calcareous earth" spread out before him in hills of sand and red clay.[69] The villages were between the hills of turquoise and inhabited by about one hundred and fifty families of migrants from Badakhshan in northern Afghanistan, home to precious mines of rubies and lapis lazuli.[70]

Fraser identified several newly opened turquoise mines and reported on the disrepair of the valuable mines of old rock. The new mines, called *taza ma'dan,* often yielded stones of a pale blue rather than the deeper, sky blue of turquoise from the old mines.[71] The "Khurooch" (Khuruj) was a relatively new mine, comprising pits dug on the side of a hill, that "had the appearance of an exhausted mine" and yielded stones "seldom of much value."[72] The Ma'dan-e Siah, or Black Mine, received its name from the color of its rock, full of veins of blue turquoise.[73] At the "Kummeree" mine, Fraser found the work being conducted in "pits dug in the grey earth," one of which had flooded and been abandoned. Stones from Kummeree could be large, but their white spots greatly lessened their value.[74] The older mines, yielding the superior old rock, meanwhile, were in a ruinous state. The "Aubee," or Blue, mine was no longer worked, even though Fraser could trace veins of blue "toorquoise matter" in its lofty roof. A "white efflorescence," which the locals referred to as *zang* (alum), covered its outer part.[75] The "Khaur-e Suffeed," or White Cave, mine was a once-extensive site that had been abandoned.[76] The "finest and largest specimens of the turquoise" were found at the celebrated "Abdool Rezakee" mine. Located near the summit of a hill, it yielded stones "exhibiting a greater variety in colour and substance" than those of any other mine.[77] But its entrance, a cleft in the side of the hill, was full of debris.

In his report, Fraser highlighted the decline of the Persian turquoise industry, expressing a developmental outlook and seeking further control over nature and its output. "From time immemorial," he wrote, the turquoise mines of Nishapur had "exclusively supplied the world with the real gem of that name."[78] But the industry had fallen into ruin, as workers did not maintain the old mines or discover sufficient new ones, limiting their operations to familiar sites and leaving behind their debris to the point of choking off existing mines. What is more, while the mines were the property of the Qajar crown, they were annually farmed out to the highest bidder. There was therefore little concern for the "improvement" of the operations, Fraser lamented, as the princes who leased the mines did not venture to speculate or invest capital in them, for fear that "rents would be raised, duties imposed, exactions of various descriptions made."[79] The princes were too uncertain of their tenure to want to make permanent developments in the mines. They sought to recover the rent, or as much of it as possible, with a minimum of trouble and thus subleased the mines to villagers, of whom Fraser wrote condescendingly and with little sympathy, "Every one lives for himself, and snatches what he can. . . . Among other objects of high promise, the turquoise mines of Nishapore are left in the hands of ignorant peasants, who have neither capital to advance nor skill to direct their operations."[80] He reported that the total rent for the mines was two thousand Khurasani tumans, equal to twenty-seven hundred pounds sterling, which was considered exorbitant. Because of this, the oldest and most valuable mine, "Abdool Rezakee," valued at seven hundred tumans, was not being worked.[81]

The writings of other travelers echoed these observations. Alexandre Chodsko noted that turquoise was no longer mined as it had been: "The Persian government never makes any explorations on its own account, and is content to lease the mines at an annual rent. . . . The most valuable stones are found amongst the *débris* of the old workings and at the bottom of shafts long since abandoned."[82] Mohan Lal, a Kashmiri *munshi* who accompanied Alexander Burnes on his Central Asian expedition, visited the turquoise mines outside Nishapur in 1831 and left an account of them in his *Travels in the Panjab, Afghanistan, and Turkistan, to Balkh, Bokhara, and Herat* (1846). About an old, spacious mine, he wrote, "The firozah here was very abundant, and ran along the wall in veins. . . . In the roof of this grotto were seen very delicate hues."[83] But the villagers (*ra'iyat*) of the Ma'dan monopolized the turquoise trade and the mines, guarding them against outsiders who might seek to claim them for the Qajar dynasty or foreign companies. Mohan Lal expressed disappointment that the local workers and proprietors of the mines were wary of "strangers" and "foreigners" and prone to "run away from their villages when anyone arrives who wishes to visit the place."[84] Although he collected some specimens and saw polished turquoise fixed to the end of sticks with sealing wax for sale, he concluded that due to the fiercely localized economy of the mines, "the produce was altogether insignificant."[85]

Fraser too surveyed the economy of the turquoise trade in the vicinity of the mines. The inhabitants of the two villages of the Ma'dan held "a complete monopoly of the labour" and the operations of the mines.[86] One hundred villagers worked them, in crews of five to ten, dividing their produce collectively and contributing their share of the rent. The largest and most valuable stones were either smuggled out by the miners or taken as the property of the sovereign. Turquoise was sold at the mines in three forms, with "the price . . . in proportion to the smallness of the risk."[87] Single stones freed from the matrix to expose the size and color of the gem but not polished or cut were the most costly. Turquoise was also sold partially embedded in rock, which left its fineness and texture a mystery and thus involved greater risk. Parcels containing four stones in this state were offered for one hundred to one hundred fifty rials, or about seven to eleven pounds sterling.[88] Turquoise was also sold in "lumps" of rock, completely hidden within the matrix. This rough form, bought in bulk, was cheapest by weight but involved the highest risk.[89]

The turquoise trade spread beyond the mines and Nishapur. The nearby Shi'i shrine city of Mashhad was home to a number of stonecutters, who polished the gem on a turning wheel of emery and gum lac mixed with sand and water. Rings and unset stones were sold in "every quarter" of Mashhad. Several of the city's caravanserais almost exclusively housed turquoise merchants, who employed stonecutters to prepare their commodity for various markets.[90] Fraser reported that the Shi'is who made pilgrimage to the shrine of Imam Riza in Mashhad purchased a great number of stones, seeking to carry "a ring of turquoise from the sacred city" as a memento (see fig. 7).[91] From Mashhad, the turquoise of Iran was traded in various directions. According to Fraser, the finest stones, from the "Abdool Rezakee" mine, met "a great sale" in Central Asia, India, and Europe, finding their way to their destinations "by the way of Herat and Candahar" and "from the Gulf of Persia."[92] Buyers in Qajar Iran and the neighboring Ottoman Empire also sought this grade of turquoise.

Fraser found prices too high at the mines, with bargains being rare, and so declined to purchase stones there or follow through on his plans to make a venture on their trade. This cost was perhaps due to the miners regarding him as a stranger or even suspecting him of being a spy sent to collect taxes. Fraser surmised that it was largely because the stone was less valued around the globe than locally, where it was believed to carry "talismanic virtue," with its native name, "Feerozah," signifying "victorious, triumphant, prosperous."[93] By the early nineteenth century, it was a commodity more dear "at the place of production, than in those where it is consumed."[94] Because turquoise had been highly esteemed for centuries at home in Iran, its merchants expected to make a great profit when they exchanged it in foreign markets.

THE OTHER SIDE OF THE WORLD

News of the demand for the riches of the earth and of the discovery of vast mineral wealth worldwide rekindled the interests of the Qajar

FIGURE 7. The Friday mosque and the shrine complex in Mashhad, a hub of the turquoise trade. Album 296 (1883), 64, Gulistan Palace Museum, Tehran.

dynasty in the turquoise mines of Nishapur. By the mid-nineteenth century, tales and reports of the finding of gold in the New World, or the "other side of the world" (Yingi Dunya), had reached Qajar court circles. These had been in circulation in the Near East and South Asia since the sixteenth century, in such Muslim accounts of discovery and conquest as *Kitab-i Bahriye* by the Ottoman sea captain Piri Reis. Even more significant in presenting a general sketch of European voyages and encounters in the New World was the anonymous Ottoman text commonly referred to as *Tarikh al-Hind al-Gharbi,* or "History of the West Indies," also called *Hadith-i Naw,* "New history."

Tarikh al-Hind al-Gharbi (1583) is a compilation of contemporary Spanish and Portuguese travel accounts—by Francisco Lopez de Gómara, Peter Martyr d'Anghiera, Augustín de Zárate, and Fernandez de Oviedo, among others—translated into Turkish and rearranged. A widely circulated version appeared from the presses of Ibrahim Muteferrika in Istanbul in 1729, and a Persian translation was produced in Mughal India.[95] *Tarikh al-Hind al-Gharbi* presents the basic narrative of Muslim histories of the New World, chronicling the first voyages of "discovery" (*inkishaf*), Columbus finding the Americas, and the subsequent conquest and creation of colonies, which brought into Europe's possession vast lands and abundant resources, including silver (*nuqra*) and gold (*tala*).

Amid the proliferation of global travel literature and geographical histories in the nineteenth century, accounts of the New World circulated widely in Asia and the Near East. In Iran, travel writing on Europe, or "Farang," came to refer to the New World—alternately called Yingi Dunya, Alam-i Naw, and Dunya-yi Naw.[96] And in 1871, the Qajar dynasty commissioned a history of the New World, titled *Tarikh-i Inkishaf-i Yingi Dunya,* based on earlier works, most notably *Tarikh al-Hind al-Gharbi.* Dedicated to Nasir al-Din Shah Qajar (r. 1848–96), it chronicles Europe's encounter with the New World in twenty-five chapters detailing the history of the ship and the compass, the journeys of Portuguese explorers around the Cape of Good Hope (Damagha-yi Umid) in the quest for the gold and spices of the Indies and their establishment of colonies from India to Brazil, Christopher Columbus's landing in the Americas in 1492, Vasco Núñez de Balboa's exploration of the "Southern Seas" and crossing of the isthmus of Panama to reach the Pacific Ocean (Muhit-i Pacific), and Ferdinand Magellan's navigation of the Pacific in search of a westward route to the Indies.[97]

The closing chapters cover the defeat and eclipse of indigenous Mesoamerican tributary empires and the European appropriation of the vast resources of the colonies in the New World. Several chapters are devoted to the conquistador Hernán Cortés's voyage to the Yucatan Peninsula, encounters with the Aztec Empire, and conquest of Mexico. According to *Tarikh-i Inkishaf-i Yingi Dunya,* the acquisition of Mexico was spurred by the Spanish quest for the gold of the Yucatan: "Mexico was a vast country with a powerful sultan by the name of Montezuma [the Aztec ruler is also referred to as *padishah,* the Persian

for "king of kings"], who ruled over many cities, possessed innumerable mines of gold and silver [ma'dan-i tala u nuqra], and commanded
large armies to guard the coast and interior of his kingdom."[98] Cortés's
maritime army brought an end to the Aztec Empire—and the fall of
Montezuma from his "throne of jewels"—laying claim to Mexico and
its rich mineral resources for the Spanish Empire.[99] A subsequent chapter covers the voyages of Francisco Pizarro and his conquest of the
Incaic Empire.[100] *Tarikh-i Inkishaf-i Yingi Dunya* ends with an account
of the Spanish discovery of the Potosí silver mines in the Andes Mountains ("Jabal And"). This set off a rush of "Farangis" to the New World
in the 1540s and 1550s. "They mined such a quantity of silver that for
three hundred years large ships equipped with cannon called 'kaleen'
[galleon] hauled silver from the New World to Spain, and it was the
silver of this mine that became the basis of the wealth of the world
[abadi-yi jahan]."[101]

Persianate accounts of the discovery of the New World, such as
Tarikh-i Inkishaf-i Yingi Dunya, trace how the possession of colonies
there and control over their mineral resources suddenly brought into
existence powerful European maritime empires of global sway. Although
the Qajar dynasty had no aspirations to a world empire and was vexed
to govern even Iran's provinces, tales of the New World and its abundant natural resources gradually led it to make more assertive efforts to
reclaim, extract, and export the mineral wealth that lay below the soil
of the "guarded domains of Iran."[102]

Since the mid-nineteenth century, news of the Americas had also
appeared in Persian newspapers and gazetteers. In the early 1850s, the
lithographed imperial gazetteer, *Ruznama-yi Vaqa'i'-yi Ittifaqiya,* ran
stories of the discovery of vast mines of gold (*tala*) in California.[103] It
reported that people from around the world, including immigrants
from China, were flocking to the state in search of gold.[104] The imperial
chronicle *Tarikh-i Muntazam-i Nasiri* also recounted that several
thousand people from China had gone to work in the mines of "Kaliforni."[105]

An 1852 edition of *Ruznama-yi Vaqa'i'-yi Ittifaqiya* links the discovery
of gold in the Sierra Nevada to American imperial expansion westward:

> The government of the United States of the northern New World [*Yingi
> Dunya-yi Shumal*] was always seeking to conquer land [*mamlikat giri*]. . . .
> First they took possession of the territory of Louisiana, and after that they
> incorporated the province of Florida into their country. In this time, the
> United States also brought the country of Texas, which had long been in the

possession of the government of Mexico, into their territory. And since gold was discovered in California [*ma'dan-i tala dar anja payda shud*], which had for some time belonged to the people of Spain [*awlad-i Ispaniyul*] in Mexico, it is now a part of the country of the northern New World.[106]

The *Tarikh-i Muntazam-i Nasiri* further details the gold rush in a section on events in "Amrika," announcing that "an engineer in the New World has unearthed a gold mine [*ma'dan-i tala'i*] on the banks of the Salarmanto [Sacramento] and San Juanik [San Joaquin] rivers in Kaliforni, a state taken from Mexico by the United States."[107] News of the discovery spread, and "from all directions people came to Kaliforni."[108]

The gold rush in California set off a global quest, with other countries scouring their territories in search of gold and other resources for extraction.[109] It also shifted the balance in global trade (*dad u sitat*). Because the abundant supply of gold unearthed in the Americas had prompted widespread speculation, *Ruznama-yi Vaqa'i'-yi Ittifaqiyi* predicted that the metal's price on the worldwide market would plummet in relation to silver and copper.[110] Along these lines, the Qajar gazetteer reported that California gold had driven down the value of the Mughal gold muhur coin among British tax collectors, as the global supply of the metal was now less scarce.[111] Brief geological accounts also appeared, such as in an 1852 edition of *Ruznama-yi Vaqa'i'-yi Ittifaqiya* on the finding of "gold earth" (*khak-i tala*) on the western coast of the Americas: "In the land of California, all the gold that lay in the surface earth had been extracted, and miners searched for gold in the earth of riverbeds and basins, in stones of white flint, and in mines dug into the ground. There is no place left that they have not dug up in their search for gold."[112]

Tales like these, of the gold rush in "Kaliforni" and earlier New World bonanzas, printed in mid-nineteenth-century Persian histories, newsletters, and gazetteers, spurred the Qajar dynasty to explore and utilize the mineral resources of Iran. Becoming conscious of a global quest for precious minerals, the Qajar state strove to recover and reclaim the once-famed turquoise mines of Nishapur.

THE PLACE OF THE RISING SUN

In the late nineteenth century, the Qajar dynasty engaged in the exploration and reclamation of natural environments throughout Iran and its borderlands. Imperial expeditions took measure of lands, peoples, and resources of the country and resulted in natural histories and geographical surveys. These were often included in lithographed (*chap sangi*) texts such as geographical histories and travel narratives describing the land of Iran and its frontiers. At the center of this information-gathering and knowledge-producing project was the Qajar imperial school in Tehran—the Dar al-Funun (House of crafts). Established in 1851 by the premier Mirza Taqi Khan Amir Kabir (1807–52), it was the first modern scientific school in Iran and offered a curriculum in sciences, geography, history, and languages to the Qajar elite. The collection of imperial cartography and ethnography became a preoccupation at the Dar al-Funun, a reflection of the nineteenth-century global interest in and outpouring of travelogues, gazetteers, and surveys of environments and peoples across the world.[113]

These subjects are represented in works attributed to one of the school's most illustrious graduates, Muhammad Hasan Khan Sani'al-Dawla I'timad al-Saltana (1843–96), Nasir al-Din Shah Qajar's "dragoman in royal attendance" and minister of publications. I'timad al-Saltana, who also was the Persian translator of *Tarikh-i Inkishaf-i Yingi Dunya*, made his first foray into historical geography and natural history with the chronicle *Mir'at al-Buldan* (Mirror of the lands), printed in four volumes between 1877 and 1879 in Tehran. However, he never finished this project, which was based on medieval Muslim geographies, most notably Yaqut al-Hamawi's thirteenth-century *Mu'jam al-Buldan*, and the numerous travel books (*safarnama*) and translations produced at the Dar al-Funun.[114]

I'timad al-Saltana's projects continued with his fourteen-hundred-page lithographed geographical history of the eastern province of Khurasan, *Matla' al-Shams, Tarikh-i Arz-i Aqdas va Mashhad-i Muqaddas, dar Tarikh va Jughrafiya-yi Mashruh-i Balad va Imakan-i Khurasan* (The place of the rising sun: History of the sacred land and sacred city of Mashhad, on the known history and geography of the lands of Khurasan). Printed between 1882 and 1884 to commemorate the second pilgrimage of Nasir al-Din Shah to the Shi'i shrine city of Mashhad, *Matla' al-Shams* is an encyclopedic record of the natural and built environments in Khurasan, detailing its rivers, mountains, deserts, mines, cities, villages, monuments, inscriptions, roads, caravanserais, mosques, and tombs.[115] The book was part of the Qajar venture to survey and reclaim its eastern borderlands, down to the turquoise below the ground. Far from a passive act of collecting information, this description of environments was an assertion of imperial control over Khurasan and its natural resources.[116]

Nasir al-Din Shah's second tour of the province thus became, among other things, the occasion for the revival of the turquoise mines of

Nishapur and the return of their operations to more direct state control. In line with this policy, Mirza 'Ali Quli Khan Mukhbir al-Dawla, a Dar al-Funun–educated Qajar prince and the minister of sciences and mines, acquired a fifteen-year lease of the mines, while Albert Houtum-Schindler (1846–1916), the Persian minister of telegraphs since 1876, was appointed the director of operations and given the task of monopolizing the turquoise trade on behalf of the state.

The description of the turquoise mines of Nishapur in the third volume of *Matla' al-Shams*, part of the Qajar venture to order nature on the eastern tracts of Iran, is based in part on information that Houtum-Schindler had gathered while working as the managing director of the mines and was written in collaboration with him. An English-language version of this meticulous survey was also communicated to the British Foreign Office through the government of India and printed in *Records of the Geological Survey of India* in 1884.[117] Houtum-Schindler's researches on Persian turquoise in fact drew extensively upon vernacular literature in the natural history genre of *javahirnama* (books of precious stones), several copies of which were in his diverse library of Persian books and manuscripts, now part of the Edward G. Browne Collection at the University of Cambridge (see fig. 8).[118] *Matla' al-Shams*, while itself substantially rooted in this genre, goes deeper into technical earth science than its more literary predecessors. Its description of turquoise provides substantial, and often quite vivid, details, gathered in 1882–83, of the new mining operations and conditions.

Matla' al-Shams made legible the lustrous stones that lay below the terra incognita of eastern Iran. Before the environment could be reclaimed, it had to be known, and such descriptions and reports were a necessary part of ventures to explore and develop the country's natural resources. *Matla' al-Shams* presents an imperial ecology that strives to take the measure of the natural resources in the Qajar province of Khurasan, as suggested by its detailed treatment of the natural history of turquoise, which includes a geographical and geological description of the valleys and caves where it was found, the different kinds of mines, the different types of turquoise, the methods of mining and cutting the stone, and its economy and trade.

Among the subjects that came within the ambit of nineteenth-century Persian print was nature. Through the Qajar encounter with the land of Khurasan, an account of the turquoise of Nishapur came to be etched in *Matla' al-Shams*, a record of the imperial quest to reclaim the mines and possess the sky-blue stone and its trade.

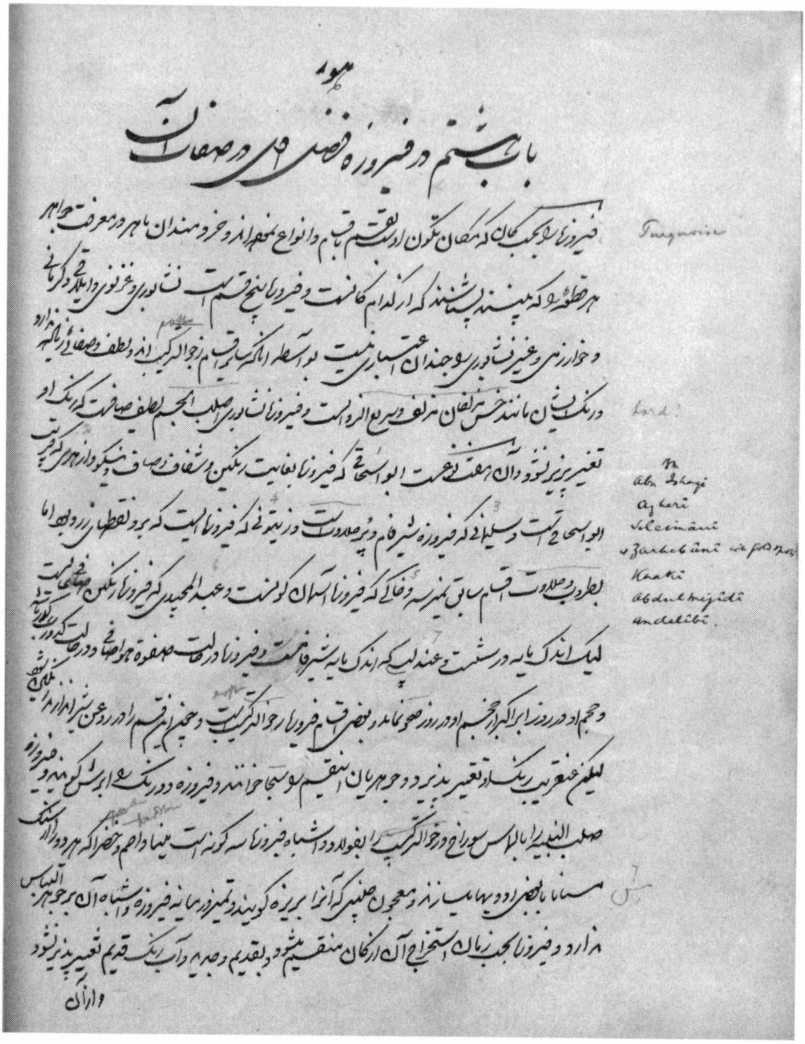

FIGURE 8. A Mongol book of stones: a description of turquoise in the Ilkhanid-era text *Tansuqnama-yi Ilkhani* by Nasir al-Din Tusi, one of the sources for Muhammad ibn Mansur's *Javahirnama*. Annotated in the handwriting of Albert Houtum-Schindler, the director of the Persian turquoise mines in 1882–83. Browne MS P. 29 (9)(1883), fol. 113, Edward G. Browne Collection—Persian Manuscripts, Cambridge University Library.

TABLE 3   THE NAMES OF THE TURQUOISE MINES OF NISHAPUR ACCORDING TO
PERSIAN BOOKS OF STONES

Book of stones	Names of the mines
Tansuqnama-yi Ilkhani (Central Asia, possibly Iran, c. 1265)	Abu Ishaqi, Azhari, Sulaymani, Zarhuni, Asuman Gun/Khaki, Kaftari, Safid Zard Fam
Javahirnama-yi Sultani (Iran, c. 1475)	Abu Ishaqi, Azhari, Sulaymani, Zarhuni, Khaki, 'Abd al-Majidi, Andalibi
Javahirnama-yi Humayuni (India, c. 1529)	Abu Ishaqi, Azhari, Sulaymani, Zarhuni, 'Abd al-Majidi, Andalibi, Asuman Gun
Matla' al-Shams (Iran, 1882–84)	'Abd al-Razzaqi (formerly known as Abu Ishaqi), Ghar-i Surkh, Ghar-i Shahpardar, Ghar-i Aqali, Maliki, Upper Zaqi, Lower Zaqi, Mirza Ahmadi, Karbala'i Karimi, Darra Kuh, 'Ali Mirza'i, Ghar-i Ra'is, Ardalani, Ghar-i Sabz, Anjiri, Kamari, Khuruj

INTO THE TURQUOISE MINES

The description of the turquoise mines of Nishapur (*ma'dan-i firuza-yi Nishapur*) in *Matla' al-Shams* follows certain themes and conventions of *javahirnama*, but much of its outlook and information—such as surveying the newly opened mines and detailing the condition of the mines of the valuable old rock—was unprecedented (see table 3). It begins with the familiar claim that turquoise is of two types: one from the mountain (*kuhi*), only separable by human labor from the rocks to which it is attached, and one from the earth (*khaki*), broken away from its rocks by the elements—rain, snow, sun, and wind—and found in the open plains at the feet of the mountains.[119]

Matla' al-Shams then goes into much deeper detail regarding the geology of the turquoise mountains of Nishapur. The mountain mines were in caves (*ghar*) in six valleys (*darra*). The most prized of these valleys was still the old 'Abd al-Razzaqi (formerly known as Abu Ishaqi), with an extensive mine 160 feet deep by the same name. However, very little turquoise was coming from there by the 1880s, even though these stones were treasured far more than those of the caves of other valleys. The valley of 'Abd al-Razzaqi also contained the Ghar-i Surkh, Ghar-i Shahpardar, and Ghar-i Aqali mines, all of which were in ruins (*matruk*).[120]

Darra-yi Safid, the White Valley, held the Maliki and the upper and lower Zaqi mines, which were once immense but had been almost entirely filled up and then abandoned in this ruined state. The vertical shafts and lateral galleries that had provided lighting and ventilation to these mines in the Safavid period now lay beneath the rubble. A number of attempts were made to clear all the mines but abandoned without result, even though there remained good turquoise to be found there. Miners (ma'danchi) were left to search for stones in the remnants and detritus of the old mines. The only working mine in the White Valley was the Mirza Ahmadi, known for good turquoise but also for precarious conditions, exacerbated by loose debris and ruined galleries.[121]

The third turquoise valley was Darra-yi Darra Kuh, the Valley of the Mountain, with the Karbala'i Karimi and Darra Kuh mines. The latter was a very deep and extensive old mine, feared by miners since loose rock had buried several workers in its tunnels (naqbha). One of its galleries was merely one to two feet wide and went down a hundred feet through loose rocks. Only three or four miners dared to work in this gallery, which was called Pul-i Sarat—"bridge in hell."[122] It is not hard to imagine the eerie and dangerous work of mining for turquoise. Returning to the surface with sky-blue stones in hand after descending into the darkness of the shaft was not assured. It had long been observed that the risky condition of these unmaintained mines provoked much fear and consternation among miners, who refused to enter them. With work halted, the mines were completely neglected, to the point that, as Jean Chardin observed in the seventeenth century, the subterranean galleries had deteriorated and "crows housed themselves in the mines."[123]

Darra-yi Siyah, the Black Valley, contained the 'Ali Mirza'i and Ghar-i Ra'is mines. Soft and brittle rocks that crumbled and filled up the tunnels made the 'Ali Mirza'i treacherous. One of its sections was fittingly called Ghar-i Birahru—"cave with no path."[124] The Ra'is mine yielded turquoise of fine color and great size, but because it was a mine of new rock, its stones faded. In 1882, a walnut-size turquoise of fine blue from this mine was gifted to Nasir al-Din Shah but lost its color after two days.[125]

Darra-yi Sabz, the Green Valley, had the Ardalani, Sabz, and Anjiri mines. The first was once an extensive mine with twelve shafts, all of which were blocked by the 1880s. What remained open was poorly ventilated, and because of the difficulties of lighting a lantern there, its galleries were called Chiraq Kush—"killers (or extinguishers) of light." But this mattered little, as the mine possessed inferior turquoise. Stones from the Ghar-i Sabz, another old mine, were green, as the mine's appellation suggested. The Anjiri was a new mine whose name referred to the fig tees in the valley. In the 1870s and 1880s, many beautifully colored (khush rang) stones were extracted from Anjiri and sold across the world, but they faded fast, causing consumers to lose trust in Persian turquoise. The impermanence of the color of these stones, along with those from the Ra'is mine, devalued the global turquoise market. Because these stones of new rock began to lose their color once dry and turned white within one to two years, some merchants resorted to keeping them wet until the time of sale. This further devalued Persian turquoise, and European jewelers agreed to purchase the stones only at very low prices.[126]

The Kamari Valley held the Kamari and the Khuruj turquoise mines. In the 1880s, neither was being worked, as water filled the former and debris blocked the latter. Houtum-Schindler and I'timad al-Saltana's survey of the various turquoise mines outside Nishapur in *Matla' al-Shams* ends with the statement "There are many more mines with names, perhaps a hundred, and more than a hundred nameless ones," an assertion of the seemingly endless prospects for finding new deposits of turquoise—and an admission of the limits of geological knowledge about the depths of the earth and its mineral resources.[127]

The most important characteristic of turquoise was its color. *Matla' al-Shams* states that "the best color of turquoise is the color of the sky [*rang-i asumani*]. . . . Not the sky that touches the horizon, where the earth and sky meet [*afaq*], but the very top of the sky [*samt al-ras*]." The color was intertwined with another, more ineffable, characteristic, referred to in the language of jewelers as the stone's *zat*, "a quality and beauty only experts could see and that words cannot explain [*az hadd-i bayan kharij ast*]."[128] Put simply, if a turquoise did not possess *zat*, it was seen as flawed. The most beautiful pieces came from the old mines (*ma'dan-i qadimi*), such as 'Abd al-Razzaqi, Khariji, Shahpardar, and Ghar-i Safid, whose stones preserved their sky blue. In the newer mines, on the other hand, "turquoises of great beauty and size were found, but their color soon fades away."[129]

Matla' al-Shams depicts the turquoise mines of Nishapur as lying in a state of ruin and far from what they had been in the reign of the Safavids. Earlier, "in the times when people worked these mines, there was order [*nazm*] and organization [*tartib*], following the standards and methods of mining [*muvafiq-i qavanin va asalib-i ma'danchigari buda*]. They dug galleries and tunnels in the rocks to allow air and light to pass into the caves, and apparently until the end of the Safavid dynasty, the state itself worked the mines [*khuda dawlat dar an ma'dan kar mikarda*]."[130] The fall of the Safavids ended the imperial monopoly on the turquoise mines, a practice that Shah 'Abbas I had introduced.

On ascending the throne, following a century of dynastic instability, the Qajars instead farmed out the mines on a yearly basis to the highest bidder, usually the governor of Khurasan. He, in turn, rented them to local villagers, who, seeking quick returns, abandoned accepted standards, cutting away at the supporting pillars and rocks between shafts and causing some of the most extensive mines to cave in.[131] Until the 1880s, the Qajar dynasty lacked the sustained interest necessary to undertake the full revival of the mining industry of Nishapur and remained content with receiving a yearly rent from the turquoise mines.

THE LABOR OF THE MINES

Many of the villagers of the Ma'dan were involved in some way in the mining, cutting, and sale of turquoise. A Qajar-era geography and survey of the Nishapur region written in 1878 divides the villages of the Ma'dan into nineteen quarters and estimates their total population as 2,151.[132] Of these, two hundred were miners (ma'danchi), and another twenty-five to thirty were "white beards" (rish safid), the local turquoise merchants.[133] While most of the mining there in the late nineteenth century was still done by hand using shovels (bilam) and pickaxes (kuland), there was an increased use of gunpowder (barut), which contributed to the deterioration of the mines by blasting away the supporting walls of galleries and tunnels.[134] After broken fragments of rock containing turquoise veins were raised to the surface of the mine, workers and merchants sorted and searched through them.

A photograph in the Gulistan Palace Museum in Tehran taken in 1894 by 'Abdallah Khan Qajar, the "Private Photographer of His Majesty the Shah," depicts a group of miners at work at the government-run Ra'is mine and accompanies a lengthy description of the extraction of turquoise (see fig. 9).[135] Having trained at the Dar al-Funun, the imperial school of sciences in Tehran, and studied lithography and photography in Paris, 'Abdallah Khan was commissioned by the Qajar court to photograph cities, shrine towns, mosques and other monuments, landscapes, and people all over Iran. In 1894 he was sent to photograph and survey the turquoise mines outside Nishapur. Below his photograph of the Ra'is mine in Darra-yi Siyah (Black valley), he describes sixty laborers, including men with pickaxes (amala va daylamchi) and hammerers (chakushzan), working diligently in the mine, overseen by a supervisor or foreman (zabitbashi). The workers entered the mine through a cave and then descended a narrowing diagonal

FIGURE 9. Hunting for turquoise at the Ghar-i Ra'is mine, Nishapur, Iran. Photograph by 'Abdallah Khan Qajar, album 291 (1894), 72, Gulistan Palace Museum, Tehran.

tunnel before scrambling down the walls to a depth of ninety feet. Going with them, 'Abdallah Khan crawled and slithered through the dark and suffocating passageways of the mine until he could "no longer draw a breath."[136] Once in the galleries, lit by lanterns of sesame oil, miners followed the veins of turquoise (*rag-i sang*) and broke the rock into rough pieces with hammers and pickaxes, which 'Abdallah Khan vividly recalled hearing deep within the mine. These pieces were then hoisted in sheepskin bags by rope and wheel (*charkh*) from the mine. Outside, some workers broke the azure stones from their rock matrix and placed them in clay pots, while others searched through the fragments for more remnants of precious turquoise. Although they had been worked for many years, 'Abdallah Khan reported that "there was still an abundance of turquoise veins in the mines" of Nishapur.[137]

The *ma'danchi* worked in often suffocating conditions for little profit, since they did not cut the stones out of the matrix and thus did not know what quality they possessed.[138] Mining was difficult and unhealthy due to the bad air and want of ventilation, worsened by blasts of gunpowder. Perhaps for this reason, it was commonly related that miners smoked opium. About a third worked the *khaki* mines, in the alluvial earth and rocks washed to the feet of mountains. The premium turquoise of the *angushtari,* or finger-ring grade, was in these mines, which were mostly worked by youths.[139] Some miners alternated between the *khaki* and the *kuhi* mines, although it was generally acknowledged that the best workmen favored the latter and sent their children to explore the *khaki* surface diggings, where the work was easier but finding turquoise was often a matter of chance.[140] Miners earned, on average, five qirans (the equivalent of five thousand dinars, or about sixteen pence) a day, in turquoise.[141] Moreover, they had to pay annual taxes to the Qajar state for the right to mine, set at sixty tumans a head (six hundred qirans) per year. Given their daily earnings, this meant a miner would break even after 120 days of working the turquoise caves. Albert Houtum-Schindler wrote that after paying this lease in turquoise, miners earned a profit of around one hundred fifty tumans (fifteen hundred qirans) per year.[142] In order to secure their livelihoods, miners smuggled stones and sought under-the-table transactions, hoping to evade taxes and their avaricious superiors.[143] The head miners or mining engineers (*ustad-i ma'danchi*) collected the turquoise for sale and rarely came away empty handed.

About twenty-five white beards acted as the local merchants and middlemen, purchasing turquoise from the head miners. Whereas the miners sold stones still embedded in the matrix and were therefore uncertain of their value, the white beards cut and sorted them for trade, making substantial profits, particularly on the individually sold *angushtari* stones, as well as an estimated 20 percent on sales to other commission agents (*dallal*) or directly to merchants or gem cutters (*hakkakan*) in Nishapur and Mashhad (see fig. 10). The latter cut the larger stones into conical and flat shapes with a rotating wheel made of emery from Afghanistan and gum from India and used rocks and pieces of sandstone for the smaller ones, an older method. After the stones were polished with leather and turquoise dust, the merchants of Mashhad inspected and sorted them for sale in Iran and abroad.[144] In addition to *angushtari,* the turquoise mined at Nishapur could be classified as *barkhana,* to be encrusted on jewelry, amulets, sword hilts, and horse trappings, or *Arabi,* taken for sale on pilgrimages to Mecca. These varieties were sold by quality and size throughout Europe, Asia, and Africa.[145]

FIGURE 10. Turquoise city: Friday mosque and panorama of Nishapur. Album 296 (1883), 14, Gulistan Palace Museum, Tehran.

The merchants of Mashhad often sent their turquoise to their contacts and commission agents in Europe, "Farangistan," where the stone was in high demand and sold for three to four times its price at the mines.[146] Europeans were never able to gain a foothold in the Persian turquoise trade or to buy stones directly at the mines outside Nishapur, where their price was far less than in Europe: I'timad al-Saltana noted that a stone bought for a few pounds at the mine where it was unearthed could be valued at hundreds more on European markets. There was once a white beard who bought a piece of turquoise from a mining engineer for less than seven tumans and then without cutting it sold it in Mashhad for ninety tumans. After a *hakkak* had shaped the stone, a merchant by the name of Mirza Hidayat Tajir shipped it to Paris, where it was appraised at fifteen hundred tumans (approximately 540 pounds).[147] *Matla' al-Shams* estimated that "in recent years," turquoise worth approximately twenty thousand tumans, or nearly seventy-two hundred pounds sterling, was purchased at the mines annually, then sold elsewhere for about sixty thousand tumans, or approximately 21,600 pounds.[148]

Domestically, in mid-nineteenth-century Iran, turquoises from Nishapur were considered gems equal to diamonds. An 1857 issue of the lithographed Qajar gazetteer *Ruznama-yi Dawlat-i 'Awliya-yi Iran* details the discovery of a flawless, radiant turquoise the size of an almond shell, comparing it to the famed Kuh-i Nur (Mountain of light) and Darya-i Nur (Sea of light) diamonds of the Mughal dynasty.[149] As scientific circles in the Qajar dynasty became increasingly concerned with the country's mineral deposits, it searched for new mines. In 1863, *Ruznama-yi Dawlat-i 'Awliya-yi Iran* reported that the director of the imperial school Dar al-Funun, 'Ali Naqi Khan I'tizad al-Saltana, had dispatched one Hajji 'Ali Akbar Amin along with a number of miners to explore newly discovered turquoise deposits near Zarand Pada in Kirman Province, where the stones were reputed to be even more radiant than those of Nishapur.[150]

THE QAJAR DYNASTY AND THE RECLAMATION OF THE TURQUOISE MINES

In the 1880s and 1890s, the Qajar dynasty assumed a developmental stance toward the turquoise of Nishapur and its worldwide trade, seeking the possession and control of the mines. This project commenced during Nasir al-Din Shah's second pilgrimage to Mashhad and continued through the writing and printing of *Matla' al-Shams,* a commemoration of his journey, whose author, I'timad al-Saltana, hoped to encourage the Qajar state in its revival of the ramshackle turquoise mining industry, including the improvement of its output on the global market.[151] Before then, the shah, much like Qajar monarchs before him, had been content to smoke tobacco from his "remarkable pure blue turquoise-encrusted water pipe," or *qalyan* (see plate 10), as his European doctor Jean-Baptiste Feuvrier noted in *Trois ans à la cour de Perse,* and to farm out the mines annually to local notables such as the governor of Khurasan for the total sum of five to eight thousand tumans, or approximately eighteen to twenty-nine hundred pounds.[152] The provincial governor, in turn, rented out most of the mines to villagers from the nearby Ma'dan.

This short-term system prevailed until 1882, when the Qajars established a state turquoise company. In an attempt to bring the mines directly under government control, the Qajar prince Mirza 'Ali Quli Khan Mukhbir al-Dawla, the royal minister for the sciences and mines, acquired a fifteen-year concession on them and the rights to a monopoly on the turquoise trade.[153] According to the terms of the lease, the shah was to receive nine thousand tumans, approximately 3,250 pounds or sixteen thousand dollars, the first year and eighteen thousand tumans, approximately sixty-five hundred pounds or thirty-two thousand dollars, annually for the remaining fourteen. Mukhbir al-Dawla sold off a two-fifths share in the company and took as partners the Qajar princes Amin al-Dawla and Nasir al-Dawla and two Tehran merchants, Hajji ʿAli Naqi and ʿAbd al-Baqi. Albert Houtum-Schindler, who had formerly served as a telegraph adviser to Mukhbir al-Dawla and as the minister of telegraphs, in which post he came to be regarded as a godsend by Nasir al-Din Shah and the Qajar administration, was appointed the director of the mines and the governor of the district of Maʿdan and investigated deposits of gold and silver for the shah. Mukhbir al-Dawla and Houtum-Schindler, who traveled to Germany together to purchase a vessel for pearl diving in the Persian Gulf and were keenly interested in Iran's natural environment, had "orders to work all the mines, mountains and *khaki,* and monopolize the whole turquoise trade for the Company."[154] European contemporaries regarded Houtum-Schindler as a talented scholar with a unique knowledge of the Persian language and of Qajar Iran.[155] His voluminous library of rare Persian manuscripts, now in the Edward G. Browne Collection at the University of Cambridge, includes various books of precious stones (*javahirnama*), which he collected while director of the mines from the libraries of bibliophile Qajar princes such as Bahman Mirza Baha ʾal-Dawla, Farhad Mirza Muʿtamad al-Dawla, and Iʿtimad al-Saltana.[156]

To establish the turquoise monopoly, Houtum-Schindler shipped all stones from the mines to the Qajar capital of Tehran. In his 1883 annual report, he recounts his efforts to reclaim the mines from villagers and to maximize the profits of the state-run company. The Qajar dynasty's seizure of the mines earned the dissatisfaction of the villagers of Maʿdan, who were now officially prohibited from engaging in the independent trade and traffic of turquoise and had effectively become laborers for the state.[157] The local merchants known as white beards lost out on the kinds of gains they had when they could purchase stones directly from the mines and sell them to the stonecutters at Mashhad. The newly imposed policies stirred unrest and resistance at the mines, where local autonomy had prevailed for nearly two centuries. Local merchants opposed the presence of a state company that outlawed their business, and when the partners neglected to supply Houtum-Schindler with money, the rank and file of miners went on strike and rebelled after not receiving their wages for seven weeks.[158]

What is more, the turquoise trade turned out to be less profitable and of smaller scale than expected. It was not long before any pretense of consolidating control over the mines was abandoned. While still seeking a monopoly on the turquoise trade, which could be conducted only through company agents, the company's partners reverted to tax-farming certain mines to locals. Houtum-Schindler quit his post in 1883 over his differences with the partners, as he resented their continual interference in the working of the mines and their neglect of finances, which undermined the entire venture. Given the persisting conditions, he found it practically impossible for the company to do more than break even, writing pessimistically, "The mines may with these arrangements be worked without a loss, but the profit, if there is any, is very small."[159] The state company's total income from and total expenditure on the turquoise mines both equaled roughly twenty thousand tumans (approximately seventy-two hundred pounds) per year.[160]

By 1890, the Qajar concession to Mukhbir al-Dawla had lapsed, and the turquoise mines were again farmed out on an annual basis. When the British MP and future viceroy of India George Nathaniel Curzon traveled through Nishapur in 1890, he found that the syndicate had leased the mines to the enterprising chief of the merchants' guild of Mashhad, the Malik al-Tujjar, who in turn had subleased them to local villagers.[161] In that year, the output of turquoise from Iran was estimated to be eighty thousand tumans, equivalent to 22,850 pounds sterling, and Curzon speculated that "these may for all practical purposes be regarded as the only mines in the world that are worked or that repay working on a large scale, and as the source of 999 out of every 1,000 turquoises that come into the market."[162] In 1894, a "local chief and a local banker" gained control over the mines.[163] In 1895, E. C. Ringler Thomson, the British vice-consul of Mashhad, reported that "Mirza Yusuf Khan, the late Afghan Agent at Mashhad," had sought to obtain the turquoise concession, which had instead been "placed under the prince Nair-ud-Dowleh, Governor of Nishapur."[164] Lessees became virtual governors of the mining district. The mines were leased at fourteen thousand tumans, approximately 5,055 pounds, and the lessees collected nine thousand tumans, approximately 3,250 pounds, in taxes from the miners and their families. The lessees worked the 'Ali Mirza'i, Zaqi, and Ra'is mines and subleased twelve others to a hundred locals for a total of 1,250 tumans, approximately 450 pounds. The goal of the lessees, as Ringler Thomson put it, was "to recoup their outlay . . . in the speediest manner."[165] In order to do so, they worked the mines without care for long-term condition. They employed approximately one hundred miners at their main mines, but their subleasers were responsible for paying the bulk of the taxes. The lessees not only overburdened the miners with crushing taxation but were notorious for not paying their laborers, who had to work the vacated and unclaimed mines of the district in their spare time to make a living.

Like other observers before him, Vice-Consul Ringler Thomson found the mines in a ruinous condition and their affairs in disarray. The old ʿAbd al-Razzaqi, or Abu Ishaqi, mine, known for its superior turquoise, had caved in completely by 1895 and was now "in the possession of one solitary individual, groping among the huge mounds of debris in the hope of turning up a valuable stone."[166] This was the last of the great old mines to collapse, after which, the vice-consul pessimistically concluded, stones of perfect color would rarely be found. A decade earlier, Houtum-Schindler had reported that fast-fading stones from the new rock mines, such as Ghar-i Raʾis, dug to replace the older ones, were in global circulation and had devalued Persian turquoise worldwide. As stones from the new mines continued to flood the market, turquoise came to be regarded as merely a semiprecious stone, an inconstant, unstable mineral compound that also lacked the hardness of diamonds, rubies, and emeralds. It earned the reputation among some as "the most treacherous of all precious stones," its fading color even associated with the personal misfortunes of the consumer.[167] The fading of turquoise, a geological phenomenon due to its chemistry and composition, became a source of folkloric beliefs and superstitions. Some thought that the mysterious manner in which turquoise could change its shade indicated the health and well-being of its wearer. And as a European talisman of love of inconstant color, it was believed to reflect the affection of its gifter, "this stone being as liable to change and caprice as the human heart itself."[168]

Ringler Thomson estimated that the mines shipped twenty-one thousand tumans of turquoise to Mashhad annually.[169] This supply was to meet demand in fin-de-siècle Asia and Europe, where rich and poor alike prized the stone. But new developments in the late nineteenth century were transforming the routes and patterns of the world turquoise trade. In Qajar Iran, the opening of the Russian Trans-Caspian Railway and the pacification of Turkmen slave raiders in the eastern borderlands more fully integrated Nishapur and Mashhad into global circuits of trade and communication. This raised the value of turquoise tenfold in Mashhad, which sent its highest-grade stones to European markets in unprecedented volume.[170] Florence's jewelers were purported to have "trunks full" of Persian stones and crowded their shops with the pale blue turquoise that Mediterraneans preferred. Persian turquoise stones

TABLE 4 PRODUCTION OF TURQUOISE IN QAJAR IRAN AND THE UNITED STATES, 1881–1895

Year	Qajar Iran (in pounds sterling)	United States (in dollars)
1881	19,580	—
1882	21,600	—
1883	21,600	2,000
1884	21,600	2,000
1885	—	3,500
1886	—	3,000
1887	23,000	2,500
1888	23,000	3,000
1889	23,000	23,675
1890	22,850	28,675
1891	—	150,000
1892	—	175,000
1893	—	143,146
1894	13,700	34,000
1895	13,700	50,000

SOURCES: Muhammad Hasan Khan Sani ʿal-Dawla Iʿtimad al-Saltana, *Matlaʿ al-Shams, Tarikh-i Arz-i Aqdas va Mashhad-i Muqaddas, dar Tarikh va Jughrafiya-yi Mashruh-i Balad va Imakan-i Khurasan* (1882–84), vol. 3 (Tehran: Farhangsara, 1986), 871; Albert Houtum-Schindler, "The Turquoise Mines of Nishapur, Khorassan," in *Records of the Geological Survey of India*, vol. 17 (Calcutta: Geological Survey, 1884), 141–42; Houtum-Schindler, "Neue Angaben über die Mineralreichthümer Persiens und Notizen über die Gegerd westlich von Zendjan," in *Jahrbuch der Kaiserlich-Königlichen Geologischen Reichsanstalt* (Vienna: K.K. Hof- und Staats-Druckerei, 1881), 177; George Nathaniel Curzon, *Persia and the Persian Question*, vol. 1 (London: Longmans, Green, 1889), 264; E.C. Ringler Thomson, "Report on the Trade and Commerce of Khorasan for the Financial Year 1895–96," Mashhad, June 1, 1896, 29–30, Foreign Office—Diplomatic and Consular Reports, Annual Reports, 1800, British National Archives, Kew Gardens; Joseph Pogue, *The Turquois: A Study of Its History, Mineralogy, Geology, Ethnology, Archaeology, Mythology, Folklore and Technology* (Washington DC, 1915), 135.

and rings could be found abundantly for sale in Mashhad, Bombay, Istanbul, Paris, and London. Throughout Qajar Iran, the stone remained a keepsake, and Ringler Thomson recalled seeing ring stones inscribed with mottoes of love: "Darling" (*nazanin*), "I am your sacrifice" (*fidayat sham*), or simply "Come" (*biya*).[171] Despite this ongoing traffic, by the last decades of the nineteenth century, the golden age of the Eurasian turquoise trade had ended. As the world's most valuable and flawlessly colored turquoise was lost with the collapse of the mines of old rock in Iran and as rich turquoise mines were opened in the New World in the 1880s and 1890s, the global patterns of the stone's trade shifted (see table 4). The Eurasian turquoise trade and its culture belonged to the age of early modern Islamic

FIGURE 11. Black and blue: the ruined Turquoise of Islam in the late nineteenth century, four hundred years after its construction. Photograph by Luigi Montabone, album 374, (1863), 18, Gulistan Palace Museum, Tehran.

tributary empires and were of the past, embodied in the ruined and desolate Turquoise of Islam mosque of Tabriz (see fig. 11). The westward expansion of the United States led to the rediscovery and reopening of the lost mines that once supplied the Aztec Empire along what came to be known as the Turquoise Trail. The exceptional stones unearthed here, of an unfading cerulean that came to rival the Persian blue, brought an end to Iran's monopoly of the world turquoise trade.

LOST AND FOUND

The rediscovery of the turquoise mines of the American Southwest, the last chapter in the nineteenth-century quest to reopen lost mines of the stone, marked the ebb of the Eurasian turquoise trade. The rising global demand for precious stones and metals spurred this pursuit to find and claim the mines of the past. Tales of the discovery of silver and gold in the New World set off a rush to explore and extract mineral resources across the world. European archaeologists and explorers in the Egyptian Sinai made the first attempts to reopen deserted turquoise mines, seeking to revive the first known veins, but their stones quickly faded and the projects ended in failure. Throughout most of the nineteenth century, Persian turquoises remained the premium grade on the global market, as they had been in the early modern period. But the most valuable mines of old rock, near Nishapur, had fallen into ruin by the 1880s, prompting the Qajar dynasty to attempt to recover and reclaim them. The Qajar scheme to reestablish an imperial monopoly on turquoise, following the Safavid model, proved largely ineffective in restoring the mines of old rock. The old Abu Ishaqi mine, with the most brilliant turquoise known in the world, lay ruined and abandoned. As newly opened Persian mines supplied the market with stones of the fast-fading new rock and the flawless sky-blue stones of Nishapur were rarely to be found, the recently uncovered turquoise mines in the American Southwest were opened up commercially for the first time, offering an alternative source of unfading blue stones that transformed the patterns and meanings of the global turquoise trade. The discovery of turquoise deposits in the New World meant that the mineral compound was not so scarce as to be found only in Iran: this substance was in rocks and strata of very similar composition on the other side of the world. Turquoise became more common, no longer a rare and peregrine stone exported from the distant East. Along the way, its material culture as a sacred object of sovereignty and imperial exchange and as the color of the urban architecture of Eurasian empires moving between the steppe and the sown faded away.

Epilogue

Indian Stone

In 1519, the ruler of the Aztec Empire, Montezuma, sent tributes of turquoise to the Spanish conquistador Hernán Cortés on the latter's arrival at Veracruz on the Gulf of Mexico. The gifts included masks, scepters, and shields of turquoise tesserae. Cortés, like other explorers of his time, had traveled to the New World in search of gold, but he found that among the Aztecs, turquoise—largely from the distant Cerrillos mines in present-day New Mexico—was more highly esteemed and held a prominent place in their regalia as an otherworldly and celestial stone.

The *General History of the Things of New Spain,* alternately referred to as the Florentine Codex, the encyclopedic sixteenth-century history of Nahuan culture compiled by the Franciscan friar Bernardino de Sahagún, depicts the moment of contact between the Aztecs and the Spanish explorer, ritualized by the exchange of turquoise. According to indigenous histories, Montezuma dispatched a party to receive Cortés with a befitting array of jewels and regalia. The most stunning of all was the mask of turquoise tesserae, believed to have been crafted in homage to the feathered serpent god Quetzalcoatl:

> This mask had, worked in the same stone, a snake folded upon itself and twisted, the bend of which was on the tip of the nose, and the twisted part went as far as the forehead and was like the bridge of the nose; then the tail and the head went in different directions, and the head with one part of the body went over one eye in such a way that it formed an eyebrow, and the tail

with part of the body went over the other eye, making another eyebrow. This mask was set in a large high crown, full of very beautiful, long, rich plumes, so that when one put the crown on one's head one also put the mask on one's face.[1]

However, as the *General History* makes clear, "the Spaniards . . . were looking for gold; they cared nothing for green-stone, precious feathers, or turquoise."[2] "Where is the gold in Mexico?" Cortés is said to have asked, and the Mexica put it all before him.[3]

Turquoise was not native to Mexico and was imported into the Yucatan Peninsula from what would become the American Southwest as tribute to the Aztec Empire. In the postclassical period (c. 900–1521), when the Mesoamerican turquoise trade reached its peak, the highest-quality stones came from the mines of Mount Chalchihuitl—given the Nahuatl name for turquoise—in the Los Cerrillos district of present-day New Mexico.[4] At the time of the Spanish arrival, the Aztec Empire spanned outward from its capital of Tenochtitlán, in the center of the Yucatan, to include much of Mesoamerica, from the Gulf of Mexico to the Pacific.[5] It had access to more raw materials through far-reaching connections of tribute and trade. Mined, cut, and polished on Mesoamerica's northern frontiers, turquoise was traded for hides, feathers, and other commodities.[6] Long-distance Mesoamerican merchants carried a range of blue- and green-hued stones into the Valley of Mexico from the mines of Cerrillos.[7]

In a return to Toltec and Mixtec traditions, the Aztecs regarded fine turquoise as a celestial mineral substance—the stone of the sky realm—and adopted it as a sacred natural object in their imperial regalia. Aztec iconography and mythology depict turquoise ornamenting the masks, shields, and diverse other objects of gods and warriors. It was associated with Quetzalcoatl and serpents of fire that descended from stars, intermediaries between earth and sky. The turquoise mosaic of a double-headed serpent and the skull of the Smoking Mirror in the collection of the British Museum in London reveal the importance of the stone in Aztec imperial symbolism and visions of nature and the universe.[8]

Mesoamerican codices and tribute rolls document the Aztec turquoise trade and its culture. Aztec codices, genealogical histories painted on paper, hide, and cloth, recount the deeds of gods and ancestors, tales of past kingdoms, religious customs, practices, and ceremonies, and calendrical cycles.[9] The Codex Mendoza, a painted history commissioned by the Spanish crown and compiled by native scribes in 1542, details the founding of Tenochtitlán and the imperial conquests of the Aztecs.[10]

It offers lists of tribute that the Aztecs received, including turquoise tesserae and masks of rich blue stone, revealing the place of turquoise in Mesoamerican trade, regalia, and customs. The Codex Mendoza also depicts Lord Eight Deer Jaguar Claw, a powerful Mixtec ruler from eleventh-century Oaxaca, receiving a ceremonial turquoise nose piercing.[11] Further references to the Aztec turquoise trade may be found in accounts of exploration in the New World. An illustration in the *General History of the Things of New Spain* shows the god Paynal, a representative of Huitzilopochtli, the deity of sun and war, in a procession, adorned with brightly colored regalia including a turquoise mosaic shield.[12] Such shields, along with jaguar skins and feathers of the quetzal, a tropical rain forest bird, formed part of the paraphernalia of Aztec warriors.

In a chapter on "the different kinds of precious stones, and how they may be sought," Sahagún details how the matrix of the "mother stone" hid turquoise and other gems:

> For there is the so-called mother. It is only a common stone, an ordinary stone; one not honored nor desirable; not regarded. Wherever it is, it is passed by; it is bypassed or just cast aside where one dwells.
>
> But this, the so-called mother of the precious stone, is not the whole thing. It is only where it is placed: perhaps well within, or on its side; not all, only a little, a bit, a small part, a fragment there wherever it is located.[13]

At sunrise, miners searched for "smoke," which they knew was a sign of precious stones and minerals:

> And those of experience, the advised, these look for it. In this manner, they know where it is: they can see that it is breathing, smoking, giving off vapor. Early at dawn, when the sun comes up, they find where to place themselves, where to stand: they face the sun. And when the sun has already come up, they are truly very attentive at looking. They look with diligence; they no longer blink; they look well. Wherever they can see that something like a little smoke column stands, that one of them is giving off vapor, this one is the precious stone. Perhaps it is a coarse stone; perhaps it is a common stone, or something smooth, or something round. They take it up; they carry it away.[14]

Here, Sahagún relates indigenous Mesoamerican geological knowledge of the mineral world, noting that it was believed among the Aztecs and their subjects that smoke and vapors rose from the mines of precious stones.

Sahagún praised the pale, smoky blue of the stone when it was removed from the mine:

And how is it with the turquoise? It comes out of a mine. From within, it is removed: the fine turquoise, the even, the smoked. . . .

It is much esteemed, because it does not appear anywhere very often. It seldom appears anywhere. This fine turquoise is much esteemed. And when it appears some distance away, it is quite pale, like the lovely cotinga [colorful bird species, sometimes blue, found in the forests of Central and South America], verily as if smoking. Some of these are flat. Some are round. . . . Some are quite smooth, some roughened, some pitted, some like volcanic rock. It becomes flat, it becomes round, it becomes pale. It smokes. The fine turquoise smokes. It becomes rough, it becomes perforated, it becomes pale.[15]

The Spanish conquest of the Aztec Empire in 1521 precipitated the decline of the Mesoamerican turquoise trade. For more than a century, they seem to have continued mining operations, but in a more oppressive and demanding imperial system. According to some accounts, the Spanish used Native American slaves in extensive operations on Mount Chalchihuitl. In 1680, a large section of a mine there suddenly collapsed, burying the miners alive and stirring an uprising that culminated in a Pueblo Indian rebellion.[16] In subsequent years, people did not dare to work the mines, for fear that they would cave in.[17] Over the next two hundred years, only the "Indians" of New Mexico visited these mines, which they did not extensively exploit, only searching around the openings and in the fragments of rock that had already been removed.[18] The main turquoise mines at Los Cerrillos were deserted, and by the mid-nineteenth century, "pine trees over a hundred years old" were growing amid their debris.[19]

At the start of that century, traces of the Aztec turquoise trade and its culture survived mostly among Native American peoples in the Pueblo region of New Mexico, including the Zuni and the Navajo, who held the blue stones in high esteem and fashioned beads and ornaments from them. Native Americans in the Southwest called turquoise *chalchihuitl,* the same name that Spanish writers had attributed to the colored stones the Aztecs prized. Native American peoples gathered these stones from Mount Chalchihuitl and other locations in the Cerrillos district of New Mexico and used them in charms, amulets, and ornaments in their ceremonial regalia and everyday lives.

Among the Zuni, a single blue turquoise bead was worth several horses. In their mythology, a blue turquoise symbolized the sky world, encompassing the upper world of the firmament and the Pacific to the west.[20] The stone had a creation myth and was personified as the legendary Hliakwa (Turquoise), who journeyed south to make his home in a high mountain.[21] In Zuni folklore, the stone carries astral significance and the radiant reflection of the Mountain of Turquoises paints the sky blue.[22] Turquoise also figured prominently as an article of exchange in Zuni networks of trade.

It is likely that the Navajo acquired the gem and the beliefs that traveled with it from the Zuni and other pueblo-dwelling tribes. The Navajo used the stone as currency and in their religious ceremonies and chants, attributing talismanic power to it and believing it to be a harbinger of good luck. Turquoise also signified the elements. Navajo threw the stone, which they believed to bring rain, into rivers as tribute to the rain god. They said that the wind blew because it was searching for turquoise. The stones were carved into the shape of horses as a means of bringing the holder into the possession of herds of the animals.[23]

Visiting Cerrillos in 1858, W.P. Blake concluded that the *chalchihuitl* of New Mexico was the precious stone turquoise after measuring the hardness and specific gravity of a specimen and comparing its reactions in front of a blowpipe to those of turquoise.[24] Moreover, they had the same composition, both being phosphates of alumina, iron, and copper. But he reported that "few or none of these stones are obtained by strangers, for they are never disposed to give for them what the Indians require."[25]

In the 1880s and 1890s, American prospectors reopened the turquoise mines in the Cerrillos district and the Burro Mountains of New Mexico, followed by others in California, Arizona, and Nevada. The Tiffany and Azure mines in New Mexico in particular produced high-quality blue stones for trade on the world market.[26] Fine sky-blue turquoise, it had been thought, could be found only in Iran, and jewelers valued only the Persian stones. By the 1890s, however, turquoise from the newly reopened American mines had taken the market and was in widespread demand, with that of New Mexico rivaling all others.[27]

The old mines on Mount Chalchihuitl were reopened following the explorations of D.C. Hyde in the 1870s but proved to be exhausted. New veins of sky blue were soon discovered nearby, however. In 1885, a Castilian mine dating from the time of the conquistadores was located, becoming known as the Palmerly claim. In 1889, the Muniz claim locked up one of the richest mines in the Cerrillos district. The Sky Blue and Morning Star mines were opened in 1891. In 1892, the American Turquoise Company of New Jersey bought all of these claims and opened the Tiffany mine.[28] Native Americans who had previously worked the mines in a noncommercial way were now outlawed from working them independently. These mines, along with a number of other claims in the Cerrillos district, produced exquisite sky-colored gems to supply the Persian blue market. New Mexico's reputation as a land of turquoise was further confirmed with the discovery of mineral veins in the Burro Mountains, most notably the Azure mine, and the formation of the Occidental and Oriental Turquoise Mining Company.

For centuries, turquoise from Iran had been in demand, with Nishapur the center of its world trade. By the late nineteenth century, however, the American Southwest was growing rapidly as a source of turquoise for the world. Stones from the Cerrillos mines of New Mexico were regarded as equally pure to Persian turquoise and were much sought after in high global demand. In 1895, the French geologist M. Carnot analyzed the physical properties of blue turquoise from Nishapur and from the Burro Mountains. Measuring traces of alumina, copper, calcium, and iron in each of the strata, Carnot concluded that despite the higher levels of iron in Persian turquoise, the two locales were fairly comparable in mineralogical properties.[29] In an age when turquoise was in fashion and consumers sought stones that would not fade, the newly formed turquoise companies of the American Southwest supplied radiant blue stones along with a guarantee that their color would last. This promise was often given in a trademark, in the form of a letter, shape, or symbol, on the back of the gem.[30] The United States Geological Survey estimated that the production of turquoise in the American Southwest had risen from a value of $3,000 in 1888 to $28,675 in 1890 and $175,000 in 1892.[31] "Indian" stones from the Americas supplied the world market. And *turquoise,* known for centuries as a stone mined in eastern Iran and named for the route of its trade to Europe through the lands of the Ottoman Turks, became the classification of stones mined by "Indians" in the American West.

. . .

Turquoise formed naturally in the depths of the earth, a phosphate of aluminum and copper. The sky blue of the stone caught the eye and led miners to descend into the depths of the earth to seek more. In early modern Islamic Eurasia, turquoise was a sacred object in the material culture and imperial regalia of tributary empires that had moved from the steppe to the sown to build metropolitan cities integrated into global networks of trade. Exchanged in imperial interactions and sent as tribute from loyal subjects to rulers, the stone and its ethereal color evolved into emblems of sovereignty.

The Islamic tributary empires of Central Asia, South Asia, and the Near East followed their initial military conquests with methods of rule based on bargaining and negotiation, which recognized the autonomy of different subjects throughout the empire. This layered sovereignty functioned through the display and projection of imperial power, supported by networks of tributary exchange. Such an indirect and mediated method of rule starkly differed from the blunt pacifying and mapping structures that formed the basis of the coercive authority of nineteenth-century colonial states. Early modern Eurasian empires, by contrast, drew authority among subjects and rival states from the representation, display, and projection of their sway and through customary and ritualized tributary exchanges. In early modern Islamic Eurasia, turquoise became treasured as a sign of earthly dominion and an object of interimperial encounters and rivalries. From there it passed to Europe as a strange and exotic stone, a marvelous possession from the faraway East, arriving in the age of discoveries and East India companies. Into the nineteenth century, the turquoise trade persisted and even expanded in volume, but its former meaning as a sacred, imperial stone faded and was forgotten. By then, European imperial access to the resources of colonies in the New World, including the turquoise mines of the Americas, had devalued the stones of Nishapur, lowering them from what they had meant as sky-blue gems and the sacred color of Islamic empires.

Notes

Unless otherwise noted, all translations are mine.

INTRODUCTION

1. *The Jahangirnama: Memoirs of Jahangir, Emperor of India,* trans. Wheeler M. Thackston (Oxford: Oxford University Press, 1999), 143. On the Safavid-Mughal rivalry over Qandahar, see Riazul Islam, *Indo-Persian Relations: A Study of the Political and Diplomatic Relations between the Mughul Empire and Iran* (Tehran: Intisharat-i Bunyad-i Farhang-i Iran, 1970).

2. In pioneering work, Jean Aubin explored the interface between pastoral nomadic and sedentary trade-based modes of circulation in the Mongol period. See, for instance, "Réseau pastoral et réseau caravanier: Les grand'routes du Khurassan à l'époque Mongole," *Le Monde Iranien et l'Islam* 1 (1971): 105–30. See also Thomas Allsen, "Mongolian Princes and Their Merchant Partners, 1200–1260," *Asia Minor* 2, no. 2 (1989): 83–125.

3. Alfred Crosby, *Ecological Imperialism: The Biological Expansion of Europe, 900–1900* (Cambridge: Cambridge University Press, 2004); Richard Grove, *Green Imperialism: Colonial Expansion, Tropical Island Edens and the Origins of Environmentalism, 1600–1860* (Cambridge: Cambridge University Press, 1995); John Richards, *The Unending Frontier: An Environmental History of the Early Modern World* (Berkeley: University of California Press, 2003).

4. For some examples of this literature, see Pamela Smith and Paula Findlen, eds., *Merchants and Marvels: Commerce, Science, and Art in Early Modern Europe* (London: Routledge, 2002); Londa Schiebinger and Claudia Swan, eds., *Colonial Botany: Science, Commerce, and Politics in the Early Modern World* (Philadelphia: University of Pennsylvania Press, 2007); Schiebinger, *Plants and Empire: Colonial Bioprospecting in the Atlantic World* (Cambridge, MA:

Harvard University Press, 2007); Harold Cook, *Matters of Exchange: Commerce, Medicine, and Science in the Dutch Golden Age* (New Haven: Yale University Press, 2007); Daniela Bleichmar, *Visible Empire: Botanical Expeditions and Visual Culture in the Hispanic Enlightenment* (Chicago: University of Chicago Press, 2012). These works share a certain outlook with an earlier strand of new history of the European Renaissance, e.g., Stephen Greenblatt, *Marvelous Possessions: The Wonder of the New World* (Chicago: University of Chicago Press, 1991); Nicholas Thomas, *Entangled Objects: Exchange, Material Culture, and Colonialism in the Pacific* (Cambridge, MA: Harvard University Press, 1991); Lisa Jardine, *Worldly Goods: A New History of the Renaissance* (New York: Doubleday, 1996).

5. Kapil Raj, *Relocating Modern Science: Circulation and the Construction of Knowledge in South Asia and Europe* (Basingstoke: Palgrave Macmillan, 2007).

6. On tributary empires, see Peter Fibiger Bang and C. A. Bayly, eds., *Tributary Empires in Global History* (Basingstoke: Palgrave Macmillan, 2011), 5–11. On concepts of Iranian kingship and sovereignty, see Abbas Amanat, *Pivot of the Universe: Nasir al-Din Shah and the Iranian Monarchy, 1831–1896* (Berkeley: University of California Press, 1997). For the South Asian context, see John Richards, *The Mughal Empire* (Cambridge: Cambridge University Press, 1996). For the Ottoman model and context, see Karen Barkey, *Empire of Difference: The Ottomans in Comparative Perspective* (Cambridge: Cambridge University Press, 2008).

7. The work closest to the sort of natural history of an object attempted here is in some ways Michael Welland's captivating *Sand: The Never-Ending Story* (Berkeley: University of California Press, 2009).

8. For some examples of the nascent literature on Islamic and Near Eastern environmental history, see Diana Davis, *Resurrecting the Granary of Rome: Environmental History and French Colonial Expansion in North Africa* (Ohio: University of Ohio Press, 2008); Richard Bulliett, *Cotton, Climate, and Camels in Early Islamic Iran: A Moment in World History* (New York: Columbia University Press, 2009); Edmund Burke III and Ken Pomeranz, eds., *The Environment and World History* (Berkeley: University of California Press, 2009); Alan Mikhail, *Nature and Empire in Ottoman Egypt* (Cambridge: Cambridge University Press, 2011); Sam White, *The Climate of Rebellion in the Early Modern Ottoman Empire* (Cambridge: Cambridge University Press, 2011); Davis and Burke, *Environmental Imaginaries of the Middle East and North Africa* (Ohio: University of Ohio Press, 2012); Mikhail, ed., *Water on Sand: Environmental Histories of the Middle East and North Africa* (Oxford: Oxford University Press, 2012).

9. Sidney Mintz, *Sweetness and Power: The Place of Sugar in Modern History* (New York: Viking, 1985). For examples of the cultural turn in commodity histories, see Timothy Brook, *Vermeer's Hat: The Seventeenth Century and the Dawn of the Global World* (New York: Bloomsbury, 2008); Sarah Stein, *Plumes: Ostrich Feathers, Jews, and a Lost World of Global Commerce* (New Haven: Yale University Press, 2008); Gary Okihiro, *Pineapple Culture: A History of the Tropical and Temperate Zones* (Berkeley: University of California Press, 2009); Robert Finlay, *The Pilgrim Art: Cultures of Porcelain in World History* (Berkeley: University of California Press, 2010). See also Arjun Appadurai, ed.,

The Social Life of Things: Commodities in Cultural Perspective (Cambridge: Cambridge University Press, 1986).

10. Thomas Allsen, *Commodity and Exchange in the Mongol Empire: A Cultural History of Islamic Textiles* (Cambridge: Cambridge University Press, 1997); Rudolph P. Matthee, *The Politics of Trade in Safavid Iran: Silk for Silver, 1600–1730* (Cambridge: Cambridge University Press, 1999). On silk, wool, and cotton textiles in Iran, see Willem Floor, *The Persian Textile Industry in Historical Perspective, 1500–1925* (Paris: L'Harmattan, 1999).

11. On spices, see Andrew Dalby, *Dangerous Tastes: The Story of Spices* (Berkeley: University of California Press, 2000); Jack Turner, *Spice: The History of a Temptation* (New York: Alfred A. Knopf, 2004); John Keay, *The Spice Route: A History* (Berkeley: University of California Press, 2006); Paul Freedman, *Out of the East: Spices and the Medieval Imagination* (New Haven: Yale University Press, 2008). For other commodities produced, consumed, and traded across Eurasia, see Rudolph P. Matthee, *The Pursuit of Pleasure: Drugs and Stimulants in Iranian History, 1500–1900* (Princeton: Princeton University Press, 2005); Finlay, *The Pilgrim Art*; Dana Sajdi, ed., *Ottoman Tulips, Ottoman Coffee: Leisure and Lifestyle in the Eighteenth Century* (London: Tauris Academic Studies, 2008); Kris Lane, *Colour of Paradise: The Emerald in the Age of Gunpowder Empires* (New Haven: Yale University Press, 2010).

12. Eurasian turquoise and its trade have not been explored as scholarly topics since nearly a century ago, with the publication of Berthold Laufer's *Notes on Turquois in the East* (Chicago: Field Museum of Natural History, 1913) and Joseph Pogue's *The Turquois: A Study of Its History, Mineralogy, Geology, Ethnology, Archaeology, Mythology, Folklore and Technology* (Washington DC: National Academy of Sciences, 1915). However, Persian turquoise also seems to have captured the interest of the late 'Abbas Iqbal of the University of Tehran, who left behind an unfinished history of jewels in Iran that was to cover the pre-Islamic and Islamic periods. For the completed section, on pre-Islamic Iran, see Iqbal, "Tarikh-i Javahir dar Iran," *Farhang-i Iran Zamin* 9 (1961): 5–45. See also M. Tosi, "The Problem of Turquoise in the Protohistoric Trade on the Iranian Plateau," *Studi di Paletnologia Paleoantropologia, Paleontologia, e Geologia del Quaternario* 2 (1974): 148, 159.

13. On this post-Timurid synthesis of the pastoral and the urban, see Maria E. Subtelny, *Timurids in Transition: Turko-Persian Politics and Acculturation in Medieval Iran* (Leiden, Netherlands: Brill, 2007). The interface between the desert and the sown pervades the writing of the fourteenth-century North African historian Ibn Khaldun; see *The Muqaddimah: An Introduction to History*, trans. Franz Rosenthal, 3 vols. (Princeton: Princeton University Press, 1958).

14. Raphaël Du Mans, *Estat de la Perse en 1660*, ed. Charles Schefer (Paris: Ernest Leroux, 1890), 192; Du Mans, *Raphaël du Mans: Missionnaire en Perse au XVIIᵉ siècle*, ed. Francis Richard (Paris: L'Harmattan, 1995), 2:149. Cited in Najaf Haider, "Precious Metal Flows and Currency Circulation in the Mughal Empire," in "Money in the Orient," special issue, *Journal of the Economic and Social History of the Orient* 39, no. 3, (1996): 298–364; Stephen Dale, *The Muslim Empires of the Ottomans, Safavids, and Mughals* (Cambridge: Cambridge University Press, 2010), 121.

15. For a programmatic essay on this theme, see David Christian, "Silk Roads or Steppe Roads? The Silk Roads in World History," *Journal of World History* 11, no. 1 (2000): 1–26. See also Anna Akasoy, Charles Burnett, and Ronit Yoeli-Tlalim, eds., *Islam and Tibet: Interactions along the Musk Routes* (London: Ashgate, 2010). For a seminal article on the possibilities of perspectives on Near Eastern history grounded in material culture, see Nikki Keddie, "Material Culture and Geography: Toward a Holistic Comparative History of the Middle East," *Comparative Studies in Society and History* 26, no. 4 (October 1984): 709–34.

16. On indigenous Eurasian cultures of exchange, tribute, and gift giving, see Finbarr B. Flood, *Objects of Translation: Material Culture and Medieval "Hindu-Muslim" Encounter* (Princeton: Princeton University Press, 2009); Linda Komaroff, *Gifts of the Sultan: The Arts of Giving at the Islamic Courts* (New Haven: Yale University Press, 2011); Komaroff, *The Gift Tradition in Islamic Art* (New Haven: Yale University Press, 2012). Among the most typical of these tributary exchanges was the presenting of *khil'at*, "robes of honor," to powerful subjects and khans in return for allegiance, but they also included other objects and resources. For a brief, informative article on the tributary economy in Iran, see Ann Lambton, " 'Piskash': Present or Tribute?," *Bulletin of the School of Oriental and African Studies* 57 (1994): 145–58.

17. Martin B. Dickson, "The Fall of the Safavi Dynasty," *Journal of the American Oriental Society* 82, no. 4 (1962): 504, 515.

18. Fernand Braudel, *The Mediterranean and the Mediterranean World in the Age of Philip II* (1966), trans. Sian Reynolds, 2 vols. (Berkeley: University of California Press, 1995).

19. The three volumes of Marshall Hodgson, *The Venture of Islam: Conscience and History in a World Civilization* (Chicago: University of Chicago Press, 1974) are *The Classical Age of Islam, The Expansion of Islam in the Middle Periods,* and *The Gunpowder Empires and Modern Times.* See also Hodgson, "The Interrelations of Societies in History," *Comparative Studies in Society and History* 5, no. 2 (1963): 227–50; Hodgson, "The Role of Islam in World History," *International Journal of Middle East Studies* 1, no. 2 (1970): 99–123. On Hodgson and world history, see Edmund Burke III, "Islamic History as World History: Marshall Hodgson, 'The Venture of Islam,'" *International Journal of Middle East Studies* 10, no. 2 (1979): 241–64; Burke, "Marshall G. S. Hodgson and the Hemispheric Interregional Approach to World History," *Journal of World History* 6, no. 2 (1995): 237–50.

20. Amin Ahmad Razi, *Haft Iqlim*, ed. Javad Fazil, 3 vols. (Tehran: 'Ali Akbar 'Ilmi, n.d.).

21. Joseph Fletcher, "Integrative History: Parallels and Interconnections in the Early Modern Period, 1500–1800," in *Studies on Chinese and Islamic Inner Asia*, ed. Beatrice Forbes Manz (London: Variorum, 1995), 3.

22. Sanjay Subrahmanyam, *Explorations in Connected History: From the Tagus to the Ganges* (Oxford: Oxford University Press, 2005); Subrahmanyam, *Explorations in Connected History: Mughals and Franks* (Oxford: Oxford University Press, 2005); Subrahmanyam, "Connected Histories: Notes towards a Reconfiguration of Early Modern Eurasia," *Modern Asian Studies* 31, no. 3 (1997): 735–62.

23. Subrahmanyam, "On the Window That Was India," *Explorations in Connected History: From the Tagus to the Ganges,* 2, 14.

24. For some recent studies that have a Eurasian frame or explore interconnections among the Safavid, Mughal, and Ottoman Empires, see Stephen Dale, *Indian Merchants and Eurasian Trade, 1600–1750* (Cambridge: Cambridge University Press, 1994); Muzaffar Alam and Sanjay Subrahmanyam, *Indo-Persian Travels in the Age of Discoveries, 1400–1800* (Cambridge: Cambridge University Press, 2007); Sebouh Aslanian, *From the Indian Ocean to the Mediterranean: The Global Trade Networks of Armenian Merchants from New Julfa* (Berkeley: University of California Press, 2011); A. Azfar Moin, *The Millennial Sovereign: Sacred Kingship and Sainthood in Islam* (New York: Columbia University Press, 2012).

25. For a critique of this view of the history of early modern Iran, see Robert McChesney, "'Barrier of Heterodoxy'?: Rethinking the Ties between Iran and Central Asia in the Seventeenth Century," in *Safavid Persia: The History and Politics of an Islamic Society,* ed. Charles Melville (London: I. B. Tauris, 1996), 231–67. For a recent work focusing on early modern Iran's foreign relations and interactions with the world, see Willem Floor and Edmund Herzig, eds., *Iran and the World in the Safavid Age* (London: I. B. Tauris, 2012).

26. Edward G. Browne, "The Persian Manuscripts of the Late Albert Houtum-Schindler," *Journal of the Royal Asiatic Society of Great Britain and Ireland* 49 (October 1917): 662.

27. C. A. Storey, *Persian Literature: A Bibliographic Survey,* vol. 2, pt. 3 (Leiden, Netherlands: Brill, 1977), 448–56.

28. Tosi, "Problem of Turquoise," 148, 159.

29. On divergence theory—the claim that access to coal, colonies, and the abundance of the New World allowed Europe to suddenly surpass Asia in the global economy circa 1800, breaking their past equivalence—see Ken Pomeranz, *The Great Divergence: China, Europe, and the Making of the World Economy* (Princeton: Princeton University Press, 2000).

1. THE COLORED EARTH

1. Joseph Pogue, *The Turquois: A Study of Its History, Mineralogy, Geology, Ethnology, Archaeology, Mythology, Folklore and Technology* (Washington DC: National Academy of Sciences, 1915), 60–67; Paul E. Desautels, *The Mineral Kingdom* (New York: Madison Square, 1968), 25.

2. Muhammad Hasan Khan Saniʿal-Dawla Iʿtimad al-Saltana, *Matlaʿ al-Shams, Tarikh-i Arz-i Aqdas va Mashhad-i Muqaddas, dar Tarikh va Jughrafiya-yi Mashruh-i Balad va Imakan-i Khurasan* (1882–84), vol. 3 (Tehran: Farhangsara, 1986), 878.

3. Desautels, *Mineral Kingdom,* 32.

4. Pogue, *Turquois,* 64.

5. William D. Neese, *Introduction to Mineralogy* (Oxford: Oxford University Press, 2000), 3.

6. Pogue, *Turquois,* 24.

7. Lina Eckenstein, *A History of Sinai* (New York: Macmillan, 1921), 30.

8. Ibid.

9. W. M. Flinders Petrie, *Researches in Sinai* (London: E. P. Dutton, 1906), 41.

10. Janine Bourriau, "The Second Intermediate Period (*c.*1650–1550 BC)," in *The Oxford History of Ancient Egypt,* ed. Ian Shaw (Oxford: Oxford University Press, 2000), 189; Shaw, "Egypt and the Outside World," in ibid., 320.

11. Jaromir Malek, "The Old Kingdom (*c.*2686–2160 BC)," in Shaw, *Oxford History of Ancient Egypt,* 105.

12. Gaston Camille Charles Maspero, *History of Egypt, Chaldea, Syria, Babylonia, and Assyria,* trans. M. L. McClure, 4 vols. (London: Grolier Society, 1903–6), 2:161–62.

13. Gae Callender, "The Middle Kingdom Renaissance (*c.*2055–1650 BC)," in Shaw, *Oxford History of Ancient Egypt,* 168.

14. Charles W. Wilson, *Picturesque Palestine: Sinai and Egypt,* vol. 4 (London: J. S. Virtue, 1881), 52.

15. Eckenstein, *History of Sinai,* 32.

16. Maspero, *History of Egypt,* 2:333.

17. Edward Henry Palmer, *The Desert of the Exodus: Journeys on Foot in the Wilderness of the Forty Years' Wanderings; Undertaken in Connexion with the Ordnance Survey of Sinai and the Palestine Exploration Fund* (Cambridge: Deighton, Bell, 1871), 233; Pogue, *Turquois,* 30.

18. Maspero, *History of Egypt,* 2:333–35; Eckstein, *History of Sinai,* 23.

19. Maspero, *History of Egypt,* 4:370.

20. Jacobus van Dijk, "The Amarna Period and the Later New Kingdom (*c.*1352–1069 BC)," in Shaw, *Oxford History of Ancient Egypt,* 307.

21. Petrie, *Researches in Sinai,* 108; Pogue, *Turquois,* 31.

22. C. E. Bosworth, "The Early Islamic History of Ghur," *Central Asiatic Journal* 6 (1961): 116–33; Richard Frye, "Firuzkuh," in *Encyclopedia of Islam,* 2nd ed., vol. 2 (Leiden, Netherlands: E. J. Brill, 1965), 928; Willem Vogelsang, *The Afghans* (London: Wiley-Blackwell, 2001), 200–203.

23. Richard Bulliet, *The Patricians of Nishapur: A Study in Medieval Islamic Social History* (Cambridge, MA: Harvard University Press, 1972).

24. For the seismic history of medieval Nishapur and a pathbreaking work on Iran's environmental history, see N. N. Ambraseys and Charles Melville, *A History of Persian Earthquakes* (Cambridge: Cambridge University Press, 1982). For earthquakes in Nishapur specifically, see Melville, "Earthquakes in the History of Nishapur," *Iran* 18 (1980): 103–20.

25. Khwaja Shams al-Din Muhammad Hafiz, *Divan-i Hafiz,* ed. Parviz Natil Khanlari, vol. 1 (Tehran: Chapkhana-yi Nil, 1980), ghazal 203, 422.

26. I'timad al-Saltana, *Matla' al-Shams,* 873.

27. Muhammad ibn Ahmad Shams al-Din al-Muqaddasi, *Ahsan al-Taqasim fi Ma'rifat al-Aqalim,* ed. M. J. de Goeje (Leiden, Netherlands: Brill, 1877), 158. See also Muhammad al-Idrisi, *Géographie d'Edrisi,* trans. P. A. Jaubert, vol. 2 (Paris: Imprimerie Royale, 1840), 185.

28. Abu Mansur 'Abd al-Malik al-Tha'alibi, *The Lata'if al-Ma'arif of Tha'alibi: The Book of Curious and Entertaining Information,* trans. C. E. Bosworth (Edinburgh: Edinburgh University Press, 1968), 131–33. For analy-

sis, see Bilal Orfali, "The Works of Abu Mansur al-Tha'alibi (350–429/961–1039)," *Journal of Arabic Literature* 40 (2009): 273–318.

29. *The Geographical Part of the Nuzhat-al-Qulub Composed by Hamd-Allah Mustawfi of Qazwin in 740 (1340)*, trans. Guy Le Strange (Leiden, Netherlands: E. J. Brill, 1919), 195.

30. Ibid., 196.

31. Abu Rayhan al-Biruni, *Kitab al-Jamahir fi Ma'rifat al-Jawahir*, trans. Hakim Muhammad Said (Islamabad: Pakistan Hijra Council, 1989), 28.

32. Ibid., 24–25, 28.

33. On lost medieval Arabic mineralogical texts, see the translator's introduction to *Arab Roots of Gemology: Ahmad ibn Yusuf al-Tifaschi's "Best Thoughts on the Best of Stones,"* trans. Samar Najm Abul Huda (Lanham, MD: Scarecrow, 1998), 3–5.

34. Ibid., 83.

35. Muhammad bin Abi Barakat Javahiri Nishapuri, *Javahirnama-yi Nizami*, ed. Iraj Afshar and Muhammad Rasul Daryagasht (Tehran: Miras-i Maktub, 2004); Nasir al-Din Tusi, *Tansuqnama-yi Ilkhani*, ed. Mudarris Razavi (Tehran: Farhang-i Iran, 1969); Nasir al-Din Tusi, "Tansuqnama-yi Ilkhani" (A.H. 973/1566), MS Browne P. 30 (8), Edward G. Browne Collection—Persian Manuscripts, Cambridge University Library; Abu'l Qasim 'Abdallah Kashani, *'Ara'is al-Javahir va Nafa'is al-Ata'ib*, ed. Afshar (Tehran: Bahman, 1966). See also Afshar, "Javaher-name-ye Nezami," in *Nasir al-Din Tusi, philosophe et savant de XIIIᵉ siècle*, ed. Z. Vesel, Afshar, and Parviz Mohebbi (Tehran: Presses Universitaires d'Iran, 2000), 151–65.

36. For instance, see Muhammad bin Abi Barakat Javahiri Nishapuri, *Javahirnama-yi Nizami*, esp. 59–64.

37. Ibid., 65.

38. Ibid., 65–67.

39. Abu Rayhan al-Biruni, *Kitab al-Jamahir*, 147–48.

40. Ibid., 148.

41. Muhammad bin Abi Barakat Javahiri Nishapuri, *Javahirnama-yi Nizami*, 127–128.

42. Nasir al-Din Tusi, *Tansuqnama-yi Ilkhani*, 76–77. See also Nasir al-Din Tusi, "Tansuqnama-yi Ilkhani," MS Browne P. 30 (8), fol. 113.

43. Nasir al-Din Tusi, *Tansuqnama-yi Ilkhani*, 79–80. See also Nasir al-Din Tusi, "Tansuqnama-yi Ilkhani," MS Browne P. 30 (8), fols. 113–14.

44. Abu Rayhan al-Biruni, *Kitab al-Jamahir*, 148.

45. Muhammad bin Abi Barakat Javahiri Nishapuri, *Javahirnama-yi Nizami*, 129–30.

46. Ibid.

47. Ibid., 131.

48. Ibid., 133.

49. Nasir al-Din Tusi, *Tansuqnama-yi Ilkhani*, 80. See also Nasir al-Din Tusi, "Tansuqnama-yi Ilkhani," MS Browne P. 30 (8), fol. 114.

50. Muhammad bin Abi Barakat Javahiri Nishapuri, *Javahirnama-yi Nizami*, 134.

51. Al-Tifaschi, *Arab Roots of Gemology*, 138.

52. Robert Finlay has made similar observations on the porcelain trade. See *The Pilgrim Art: Cultures of Porcelain in World History* (Berkeley: University of California Press, 2010), 6.

53. On early modern empires and their transformation of environments, see John Richards, *The Unending Frontier: An Environmental History of the Early Modern World* (Berkeley: University of California Press, 2003).

2. TURQUOISE, TRADE, AND EMPIRE

1. See Thomas Allsen, *Commodity and Exchange in the Mongol Empire: A Cultural History of Islamic Textiles* (Cambridge: Cambridge University Press, 1997).

2. See Rudolph P. Matthee, *The Politics of Trade in Safavid Iran: Silk for Silver, 1600–1730* (Cambridge: Cambridge University Press, 1999), 4.

3. On the consumption of medicinal plants, see Rudi Matthee, *The Pursuit of Pleasure: Drugs and Stimulants in Iranian History, 1500–1900* (Princeton: Princeton University Press, 2005); on Persian cobalt and Chinese porcelain, see Robert Finlay, *The Pilgrim Art: Cultures of Porcelain in World History* (Berkeley: University of California Press, 2010); on Colombian emeralds in Mughal India, see Kris Lane, *Colour of Paradise: The Emerald in the Age of Gunpowder Empires* (New Haven: Yale University Press, 2010).

4. On the traditions and objects of gifts and gift giving in Islamic history, see Finbarr B. Flood, *Objects of Translation: Material Culture and Medieval "Hindu-Muslim" Encounter* (Princeton: Princeton University Press, 2009); Linda Komaroff, *Gifts of the Sultan: The Arts of Giving at the Islamic Courts* (New Haven: Yale University Press, 2011); Komaroff, *The Gift Tradition in Islamic Art* (New Haven: Yale University Press, 2012). On the gift economy more broadly, see Marcel Mauss, *The Gift: The Form and Reason for Exchange in Archaic Societies,* translation of *Essai sur le Don: Forme et raison de l'échange dans les sociétés archaïques* (Paris: Presses Universitaires, 1950) by W. D. Walls (London: Routledge, 1990), 1–18.

5. See Peter Jackson and Laurence Lockhart, eds., *The Timurid and Safavid Periods,* vol. 6 of *The Cambridge History of Iran* (Cambridge: Cambridge University Press, 1986); Beatrice Manz, *The Rise and Rule of Tamerlane* (Cambridge: Cambridge University Press, 1999); Manz, *Power, Politics, and Religion in Timurid Iran* (Cambridge: Cambridge University Press, 2007); Maria Subtelny, *Timurids in Transition: Turko-Persian Politics and Acculturation in Medieval Iran* (Leiden, Netherlands: Brill, 2007); Nicola Di Cosmo, Allen Frank, and Peter Golden, eds., *The Cambridge History of Inner Asia: The Chinggisid Age* (Cambridge: Cambridge University Press, 2009).

6. Hafiz-i Abru, *Jughrafiya-yi Hafiz-i Abru: Qismat-i Ruba'-i Khurasan, Herat,* ed. Mayil Heravi (Tehran: Bunyad-i Farhangi, 1970), 64.

7. Ibid., 63–64.

8. See the brief but eloquent references to the turquoise of Nishapur in Ellen Melloy's acclaimed *The Anthropology of Turquoise: Reflections on Desert, Sea, Stone, and Sky* (New York: Vintage, 2002), 107–8. On the history of Nishapur, see Firaydun Girayili, *Nishabur: Shahr-i Firuza* (Mashhad:

Firdawsi University, 1978). For contemporary accounts of late nineteenth-century Nishapur and its society, see Muhammad Hasan Khan Sani' al-Dawla I'timad al-Saltana, *Matla' al-Shams, Tarikh-i Arz-i Aqdas va Mashhad-i Muqaddas, dar Tarikh va Jughrafiya-yi Mashruh-i Balad va Imakan-i Khurasan* (1882–84), 3 vols. (Tehran: Farhangsara, 1986), 3:822–974; Mirza Husayn bin 'Abd al-Karim Durrudi, *Kitabcha-yi Nishabur: Guzarish-i Rustha-yi Nishapur dar Sal-i 1296 Qamari,* ed. Rasul Ja'fariyan (Mashhad: Astan-i Quds-i Razavi, 2003).

9. Zayn al-Din Muhammad Jami, "Mukhtasar dar bayan-i shinakhtan-i javahir" (A.H. 1259/1843), MS Browne P. 32 (9) 2, fols. 57a–73b, Edward G. Browne Collection—Persian Manuscripts, Cambridge University Library. For an edited text of this work based on an unsigned manuscript from a collection in Iran, see "Javahirnama," ed. Taqi Binesh, *Farhang-i Iran Zamin* 12 (1964): 273–97. Binesh, however, deems it anonymous.

10. Zayn al-Din Muhammad Jami, "Mukhtasar dar bayan-i shinakhtan-i javahir," MS Browne P. 32 (9) 2, fol. 67a.

11. Ibid., fol. 67b.

12. Ibid., fol. 67a.

13. Ibid., fol. 67b.

14. For the standard account of these dynasties, see John E. Woods, *The Aqquyunlu: Clan, Confederation, and Empire* (Salt Lake City: University of Utah Press, 1999). For a Persian chronicle of the Aq Quyunlu period, see Fadlullah Ruzbihan Khunji-Isfahani, *Tarikh-i 'Alamara-yi Amini,* ed. Woods, trans. Vladimir Minorsky (London: Royal Asiatic Society, 1992); *Persia A.D. 1478–1490: An Abridged Translation of Fadlullah b. Ruzbihan Khunji's Tarikh-i 'alam-ara-yi Amini,* trans. Minorsky (London: Royal Asiatic Society, 1957).

15. Sheila S. Blair and Jonathan Bloom, *The Art and Architecture of Islam, 1250–1800* (New Haven: Yale University Press, 1994), 50.

16. Ibid., 52.

17. Manuscripts may be found in the Majlis Library in Tehran and among the Edward G. Browne Collection. In the Majlis Library Archives (Kitabkhana-yi Majlis; henceforth ML), see Muhammad ibn Mansur's "Javahirnama" MSS 9103 (n.d.), 2166 (A.H. 1230/1815), and 15062 (n.d.); "Risala-yi Iskandariya va Javahirnama" (n.d.), MS 5690/3; "Haqidat al-Javahir" (n.d.), MS 5684; "Kitab dar Javahirat va Filizat va Khass-i Har Yak" (A.H. 1283/1867), MS 2167. At Cambridge, see Muhammad ibn Mansur's "Javahirnama" MSS Browne P. 32 (9) (A.H. 1259/1843), Browne P. 31 (9) (A.H. 1260/1844), and Browne P. 29 (9) (A.H. 1300/1883).

18. Muhammad ibn Mansur, "Javahirnama," ML MS 2166, fols. 1–4.

19. Muhammad ibn Mansur, "Javahirnama," ML MSS 9103, fol. 1a, and 2166, fol. 1a.

20. Muhammad ibn Mansur, "Javahirnama," ML MS 2166, fol. 13.

21. 'Ala' ibn al-Husayn al-Bayhaqi, *Ma'din al-Nawadir fi Ma'rifat al-Jawahir* (Kuwait: Dar al-'Arabiya li al-Nashr wa'l-Tawzi', 1985), 41.

22. Ibid., 42–43.

23. Muhammad ibn Mansur, "Javahirnama," ML MS 2166, fols. 8–10.

24. Muhammad bin Abi Barakat Javahiri Nishapuri, *Javahirnama-yi Nizami* (1196), ed. Iraj Afshar and Muhammad Rasul Daryagasht (Tehran: Miras-i Maktub, 2004), 69–70.

25. Muhammad ibn Mansur, "Javahirnama," ML MS 2166, fols. 1–8.

26. On the genres of *'aja'ib* and the strange and wonderful in Arabic and Persian literature, see Roy Mottahedeh, "*'Aja'ib* in *The Thousand and One Nights*," in *"The Thousand and One Nights" in Arabic Literature and Society,* ed. Richard Hovanissian and Georges Sabagh (Cambridge: Cambridge University Press, 1997), 29–39.

27. Muhammad ibn Mansur, "Javahirnama," ML MS 2166, fols. 95–96.

28. Ibid., fols. 96–98.

29. Ibid., fols. 97–98.

30. Ibid., fols. 105–6.

31. Ibid., fol. 98.

32. Ibid., fol. 105.

33. Ibid., fols. 105–6.

34. Ibid., fol. 107.

35. 'Ala' ibn al-Husayn al-Bayhaqi, *Ma'din al-Nawadir fi Ma'rifat al-Jawahir,* 94–95.

36. For a general view of money in the Safavid economy, see Rudi Matthee, *Persia in Crisis: Safavid Decline and the Fall of Isfahan* (London: I.B. Tauris, 2012), 76–80. On metals, mints, and the monetary history of the Mughal Empire, see John Richards, ed., *The Imperial Monetary System of Mughal India* (Delhi: Oxford University Press, 1987). On the Ottomans, see Şevket Pamuk, *A Monetary History of the Ottoman Empire* (Cambridge: Cambridge University Press, 2000). For a comparative perspective, see Sanjay Subrahmanyam, "Precious Metal Flows and Prices in Western and Southern Asia, 1500–1750," *Studies in History* 7, no. 1 (1991): 79–105.

37. Willem Floor, *The Economy of Safavid Persia* (Wiesbaden: Ludwig Reichert, 2000), 303–4.

38. See, for instance, James Allan, "Early Safavid Metalwork," in *Hunt for Paradise: Court Arts of Safavid Iran, 1501–1576,* edited by Jon Thompson and Sheila R. Canby (Milan: Skira, 2003), 203–40.

39. For some brief accounts of mining in the Safavid Empire, see Muhammad Ibrahim Bastani Parizi, *Siyasat va Iqtisad-i 'Asr-i Safavi* (Tehran: Intisharat-i Safi 'Ali Shah, 1999), 219; Floor, *Economy of Safavid Persia,* 305–7.

40. Iskandar Bayg Munshi, *Tarikh-i 'Alamara-yi 'Abbasi,* ed. Iraj Afshar, vol. 1 (Tehran: Amir Kabir, 2003), 321. English translation in Eskandar Beg Monshi, *History of Shah 'Abbas the Great,* trans. Roger M. Savory, vol. 1 (Boulder, CO: Westview, 1978), 455.

41. Budaq Munshi Qazvini, *Kitab-i Javahir al-Akhbar,* ed. Muhsin Bahram Nijhad (Tehran: Ayina-yi Miras, 1999), 189.

42. Iskandar Bayg Munshi, *Tarikh-i 'Alamara-yi 'Abbasi,* 321. English translation in Eskandar Beg Monshi, *History of Shah 'Abbas,* 455.

43. *A Chronicle of the Carmelites in Persia and the Papal Mission of the XVIIth and XVIIIth Centuries,* vol. 1 (London: Eyre and Spottiswoode, 1939), 198. Cited in Floor, *Economy of Safavid Persia,* 304.

44. Jean Baptiste Tavernier, *Travels in India: Translated from the Original French Edition of 1676*, trans. V. Ball (London: Macmillan, 1889), 103–4.

45. *Arakel of Tabriz: The History of Vardapet Arakel of Tabriz*, trans. George Bournoutian, vol. 2 (Costa Mesa, CA: Mazda, 2006), 454. On the Armenian merchants of Julfa, Isfahan, and their long-distance trading networks, including references to ones dealing in jewels, see Sebouh Aslanian, *From the Indian Ocean to the Mediterranean: The Global Trade Networks of Armenian Merchants from New Julfa* (Berkeley: University of California Press, 2011).

46. *Arakel of Tabriz*, 465.

47. Ibid., 464–65.

48. Floor, *Economy of Safavid Persia*, 304. See also Mohammad Rafi' al-Din Ansari Mostowfi al-Mamalek, *Dastur al-Moluk: A Safavid State Manual*, trans. Floor and Mohammad H. Faghfoory (Costa Mesa, CA: Mazda, 2007), 104n167.

49. Mostowfi al-Mamalek, *Dastur al-Moluk: A Safavid State Manual*, 104.

50. *Tadhkirat al-Muluk*, ed. Muhammad Dabirsiyaqi (Tehran: Amir Kabir, 1989), 133–36; *Tadhkirat al-Muluk: A Manual of Safavid Administration (circa 1137/1725)*, trans. V. Minorsky (Cambridge: Cambridge University Press, 1943), 177.

51. Mostowfi al-Mamalek, *Dastur al-Moluk: A Safavid State Manual*, 113, 253.

52. Muhammad ibn Mansur, "Javahirnama," ML MS 2166, fols. 105–6.

53. Flood, *Objects of Translation*, 124.

54. Khwandamir, *Habibu's-Siyar*, vol. 3, *The Reign of the Mongol and the Turk*, trans. W.M. Thackston (Cambridge, MA: Harvard University Press, 1994), 606.

55. Ibid., 589.

56. See Cengiz Koseoglu, *The Topkapi Saray Museum: The Treasury*, trans. and ed. J.M. Rogers (Boston: Little, Brown, 1987), 17.

57. Jean Baptiste Tavernier, *Collections of Travels through Turkey into Persia, and the East Indies*, vol. 1 (London: Moses Pitt, 1684), 46.

58. Muhammad ibn al-Mubarak Qazvini, "Risala dar Ma'rifat-i Javahir" (A.H. 1299/1883), Browne P. 29 (9) 2.

59. Ibid., fol. 42a.

60. Ibid., fols. 41a–41b.

61. Koseoglu, *Topkapi Saray Museum*, 38.

62. See ibid., 40.

63. Evliya Çelebi, *Narrative of Travels in Europe, Asia, and Africa in the Seventeenth Century*, trans. Joseph von Hammer, vol. 1, pt. 2 (London: Parbury, Allen, 1834), 189.

64. *An Ottoman Traveller: Selections from the Book of Travels of Evliya Çelebi*, trans. Robert Dankoff and Sooyong Kim (London: Eland, 2010), 14.

65. Evliya Çelebi, *Narrative of Travels*, 124.

66. Evliya Çelebi, *Ottoman Traveller*, 59.

67. Evliya Çelebi, *Narrative of Travels*, 44.

68. Evliya Çelebi, *Ottoman Traveller*, 236.

69. Ibid., 299.

70. Zahir al-Din Babur, *The Baburnama: Memoirs of Babur, Prince and Emperor*, trans. Wheeler M. Thackston (New York: Random House, 2002), 6.

71. Gulbadan Begam, *The History of Humayun (Humayn-Nama)*, trans. Annette S. Beveridge (London: Royal Asiatic Society, 1902), 95.

72. The Oriental and India Office of the British Library has a manuscript of the *Javahirnama-yi Humayuni*—Muhammad ibn Ashraf Rustamdari, "Nuskha-yi Javahirnama-yi Humayuni," MS Or. 1717—dated A.H. 1268/1852. The preface is in fols. 3a–3b.

73. The precious stones that it describes are pearl (*lulu*), ibid., fol. 4b.; hyacinth (*yaqut*), fol. 11a; balas ruby (*la'l*), fol. 15b; emerald (*zumurrud*), fol. 19a; topaz (*zabarjad*), fol. 21b; diamond (*almas*), fol. 22b; cat's-eye (*'ayn al-hir*), fol. 26b; turquoise (*firuza*), fol. 27b; bezoar (*pazahr*) and other stones of animal origin, fol. 31b; carnelian (*'aqiq*), fol. 42b; ruby and sapphire (*yaqut*), fol. 43b; agate and onyx (*jaz'*), fol. 45a; magnet (*maghnatis*), fol. 45b; emery (*sunbada*), fol. 48b; lapis lazuli (*lajvard*), fol. 50a; coral (*basud* or *marjan*), fol. 53a; jasper (*yashb*), fol. 55b; and crystal (*bulur*), fol. 57b.

74. Ibid., fols. 29a, 31a.

75. *The Akbar Nama of Abu-l-Fazl: History of the Reign of Akbar Including an Account of His Predecessors*, trans. H. Beveridge, vol. 1 (Calcutta: Asiatic Society of Bengal, 1902), 435.

76. *The Shah Jahan Nama of 'Inayat Khan: An Abridged History of the Mughal Emperor Shah Jahan, Compiled by His Royal Librarian*, ed. W. E. Begley and Z. A. Desai, trans. A. R. Fuller (Oxford: Oxford University Press, 1990), 147, 219.

77. Ibid., 219–20, 495.

78. *The Embassy of Sir Thomas Roe to the Embassy of the Great Mogul, 1615–1619, as Narrated in His Journal and Correspondence*, vol. 2 (London: Hakluyt Society, 1899), 412.

79. Ibid., 480.

80. Ibid., 485.

81. Afonso de Albuquerque, *The Commentaries of the Great Afonso Dalboquerque, Second Viceroy of India*, trans. Walter De Gray Birch, vol. 4 (London: Hakluyt Society, 1875), 83, 88.

82. H. R. Roemer, "The Safavid Period," in Jackson and Lockhart, *Timurid and Safavid Periods*, 308.

83. Berthold Laufer, *Notes on Turquoise in the East* (Chicago: Field Museum of Natural History, 1913), 14–15, 20.

84. A. Campbell, "Notes on Eastern Tibet," *Journal of the Asiatic Society of Bengal* 21 (1855): 215–328, cited in ibid., 16.

85. Laufer, *Notes on Turquoise*, 18–19.

86. Ibid., 14, 18–20, 216.

87. Osvaldo Roero, *Ricordi dei Viaggi al Cashemir, Piccolo e Medio Tibet e Turkestan*, vol. 3 (Turin: Camilla E. Bertolero, 1881), 72.

88. Laufer, *Notes on Turquoise*, 13–15, 42. Muhammad ibn Mansur also notes that in Asia, turquoise mixed with other stones in the compound known

as *tar malh* was used to adorn Buddhist idols. See Muhammad ibn Mansur, "Javahirnama," ML MS 2166, fol. 104.

89. For references to the changing course of the Oxus, see Ebülgazi Bahadir Han, *Histoire des Mogols et des Tartares*, trans. Baron Desmaisons, 2 vols. (St. Petersburg: Imprimerie de l'Académie Impérial des Sciences, 1871–74), 1:221, 312; 2:207, 291.

90. J.P. Ferrier, *Caravan Journeys and Wanderings in Persia, Afganistan, Turkistan, and Beloochistan*, trans. William Jesse (London: John Murray, 1856), 94–95.

91. Amin Guli, *Tarikh-i Siyasi va Ijtima'i-yi Turkmanha* (Tehran: Nashr-i 'Ilm, 1987), 308–14.

92. For turquoise in the material culture of the Turkmen, see Ernst Johannes Kläy and Anne Brechbühl, *Islamic Collection Henri Moser Charlottenfels* (Bern: Musée d'histoire de Berne, 1991); L. Beresneva, *The Decorative and Applied Art of Turkmenia* (Leningrad: Aurora, 1976).

93. Riazul Islam, *Indo-Persian Relations: A Study of the Political and Diplomatic Relations between the Mughul Empire and Iran* (Tehran: Intisharat-i Bunyad-i Farhang-i Iran, 1970), 14.

94. Ibid., 15–18.

95. Ibid., 23–25.

96. Ibid., 41–44, 57–61.

97. *The Jahangirnama: Memoirs of Jahangir, Emperor of India*, trans. Wheeler M. Thackston (Oxford: Oxford University Press, 1999), 66–67; Islam, *Indo-Persian Relations*, 64–69.

98. Jahangir, *Jahangirnama*, 122–23.

99. Ibid., 66–67.

100. Ibid., 143.

101. Islam, *Indo-Persian Relations*, 80.

102. Jahangir, *Jahangirnama*, 384–85; Islam, *Indo-Persian Relations*, 80–82.

103. On the fall of the Safavid Empire, see Matthee, *Persia in Crisis*, 198–241; Laurence Lockhart, *The Fall of the Safavi Dynasty and the Afghan Occupation of Persia* (Cambridge: Cambridge University Press, 1958).

3. THE TURQUOISE OF ISLAM

1. For a programmatic essay on architecture and the projection of power by the Ottomans, the Safavids, and the Mughals, see Gulru Necipoglu, "Framing the Gaze in Ottoman, Safavid, and Mughal Palaces," *Ars Orientalis* 23 (1993): 303–42. For works on Ottoman architectural history, see Necipoglu, *The Age of Sinan: Architectural Culture in the Ottoman Empire* (Princeton: Princeton University Press, 2005); Shirine Hamadeh, *The City's Pleasures: Istanbul in the Eighteenth Century* (Seattle: University of Washington Press, 2007); Zeynep Çelik, *Empire, Architecture, and the City: French-Ottoman Encounters, 1830–1914* (Seattle: University of Washington Press, 2008); Çelik, Julia Clancy-Smith, and Francis Terpak, eds., *Walls of Algiers: Narratives of the City through Text and Image* (Seattle: University of Washington Press, 2009). On the

Safavids, see Stephen Blake, *Half of the World: The Social Architecture of Safavid Isfahan* (Costa Mesa, CA: Mazda, 1999); Sussan Babaie, *Isfahan and Its Palaces: Statecraft, Shi'ism, and the Architecture of Conviviality in Early Modern Iran* (Edinburgh: Edinburgh University Press, 2008); Kishwar Rizvi, *The Safavid Dynastic Shrine: Architecture, Religion, and Power in Early Modern Iran* (London: Tauris Academic Studies, 2011). On Mughal architectural history, see Thomas Metcalf, *An Imperial Vision: Indian Architecture and Britain's Raj* (Berkeley: University of California Press, 1989); Catherine Asher, *Architecture of Mughal India* (Cambridge: Cambridge University Press, 1992); Blake, *Shahjahanabad: The Sovereign City in Mughal India, 1639–1739* (Cambridge: Cambridge University Press, 2002); Alka Patel, *Building Communities in Gujarat: Architecture and Society during the Twelfth through Fourteenth Centuries* (Leiden, Netherlands: Brill, 2004).

2. On color in Islamic history, see Sheila S. Blair and Jonathan Bloom, *And Diverse Are Their Hues: Color in Islamic Art and Culture* (New Haven: Yale University Press, 2011).

3. For some architectural histories of the Timurid Empire and its heritage in Central Asia, Iran, and India, see Lisa Golombek and Donald Wilber, *The Timurid Architecture of Iran and Turan*, 2 vols. (Princeton: Princeton University Press, 1988); Sheila S. Blair and Jonathan Bloom, *The Art and Architecture of Islam, 1250–1800* (New Haven: Yale University Press, 1994), 37–54.

4. Abbas Iqbal, "Tarikh-i Javahir dar Iran," *Farhang-i Iran Zamin* 9 (1961), 36.

5. Ibid., 38.

6. Abolqasem Ferdowsi, *Shahnameh: The Persian Book of Kings,* trans. Dick Davis (New York, 2006), 28–62.

7. Ibid., 172.

8. Ibid., 332.

9. Ibid., 219.

10. Richard Eaton, *A Social History of the Deccan, 1300–1761: Eight Indian Lives* (Cambridge: Cambridge University Press, 2005), 50.

11. Mahomed Kasim Ferishta, *History of the Rise of Mahomedan Power in India,* trans. John Briggs, vol. 2 (London: Longman, Rees, Orme, Brown, and Green, 1829), 189–90; George Mitchell and Richard Eaton, *Firuzabad: Palace City of the Deccan* (Oxford: Oxford University Press, 1992).

12. Abu'l Qasim 'Abdallah Kashani, *'Ara'is al-Javahir va Nafa'is al-Ata'ib,* ed. Iraj Afshar (Tehran: Bahman, 1966). For a translation of the chapter on ceramics, see James W. Allan, "Abu'l-Qasim's Treatise on Ceramics," *Iran* 11 (1973): 111–20.

13. Allan, "Abu'l-Qasim's Treatise on Ceramics," 111–12.

14. Ibid., 114.

15. Golombek and Wilber, *Timurid Architecture,* 1:75.

16. Ibid., 91–93, 117.

17. Kamal al-Din 'Abd al-Razzaq Samarqandi, *Matla' al-Sa'dayn va Majma' al-Bahrayn,* ed. 'Abd al-Husayn Nava'i, 2 vols. (Tehran: 'Ulum-i Insani va Mutala'at-i Farhangi, 2004). On 'Abd al-Razzaq's travels and travel writing, see Muzaffar Alam and Sanjay Subrahmanyam, *Indo-Persian Travels in the Age*

of Discoveries, 1400–1800 (Cambridge: Cambridge University Press, 2007), 45–92.

18. Sharaf al-Din 'Ali Yazdi, *Zafarnama: Tarikh-i 'Umumi-yi Mufassil-i Iran dar Dawra-yi Timuriyan,* ed. Muhammad 'Abbasi, 2 vols. (Tehran: Amir Kabir, 1957), 1:781. For 'Abd al-Razzaq's references to Shahr-i Sabz in the early years of Timur's reign, see Kamal al-Din 'Abd al-Razzaq Samarqandi, *Matla' al-Sa'dayn,* vol. 1, pt. 1, 309–10.

19. Kamal al-Din 'Abd al-Razzaq Samarqandi, *Matla' al-Sa'dayn,* vol. 1, pt. 2, 809. See also Golombek and Wilber, *Timurid Architecture,* 1:258, translation of Sharaf al-Din 'Ali Yazdi, *Zafarnama,* 2:144.

20. Kamal al-Din 'Abd al-Razzaq Samarqandi, *Matla' al-Sa'dayn,* vol. 1, pt. 2, 810. See also Golombek and Wilber, *Timurid Architecture,* 1:258–259.

21. Sharaf al-Din 'Ali Yazdi, *Zafarnama* (c. 1480), fols. 359v–360r, John Work Garrett Collection, Johns Hopkins University.

22. Kamal al-Din 'Abd al-Razzaq Samarqandi, *Matla' al-Sa'dayn,* vol. 1, pt. 2, 809.

23. Golombek and Wilber, *Timurid Architecture,* 1:259.

24. Ibid., 189–93.

25. Kamal al-Din 'Abd al-Razzaq Samarqandi, *Matla' al-Sa'dayn,* vol. 2, pt. 1, 370–71.

26. Amin Sayyid 'Abdallah al-Husayni Ma'ruf bi Asil al-Din Va'iz Haravi, "Maqsad al-Iqbal-i Sultaniyya," in *Risala-yi Mazarat-i Herat,* ed. Fikri Saljuqi (Kabul: Publishing Institute, 1967), vol. 1, pt. 1, 28–29. See also Golombek and Wilber, *Timurid Architecture,* 1:307.

27. Kamal al-Din 'Abd al-Razzaq Samarqandi, *Matla' al-Sa'dayn,* vol. 2, pt. 1, 369–71.

28. Following the Russian annexation of the Afghan oasis of Panjdih in 1885, amid the threat of a Russian invasion of Herat, British engineers advised 'Abd al-Rahman, the "Iron Amir" of Afghanistan, to level the madrassa and the mosque in the Musalla to make preparations for the defense of the city. See C. E. Yate, *Northern Afghanistan: Or Letters from the Afghan Boundary Commission* (Edinburgh: William Blackwood and Sons, 1888), 65.

29. On the construction of the mosque and the madrassa in the Musalla complex, see Hafiz-i Abru, *Jughrafiya-yi Hafiz-i Abru: Qismat-i Ruba'-i Khurasan, Herat,* ed. Mayil Haravi (Tehran: Bunyad-i Farhangi, 1970), 87.

30. Kamal al-Din 'Abd al-Razzaq Samarqandi, *Matla' al-Sa'dayn,* vol. 2, pt. 1, 424.

31. On the building of the Mashhad Masjid-i Jami', see Hafiz-i Abru, *Zubdat al-Tavarikh,* ed. Kamal Hajj Sayyid Javadi, vol. 4 (Tehran: Sazman-i Chap va Intisharat-i Vizarat-i Farhang va Irshad-i Islami, 2001), 792–93.

32. Kamal al-Din 'Abd al-Razzaq Samarqandi, *Matla' al-Sa'dayn,* vol. 2, pt. 1, 821.

33. The inscriptions of the Gawhar Shad Mosque are recorded in Muhammad Hasan Khan Sani'al-Dawla I'timad al-Saltana's natural and geographical history of Mashhad, *Matla' al-Shams, Tarikh-i Arz-i Aqdas va Mashhad-i Muqaddas, dar Tarikh va Jughrafiya-yi Mashruh-i Balad va Imakan-i Khurasan,* printed between 1882 and 1884 to commemorate Nasir al-Din Shah's second

pilgrimage to the shrine of Imam Riza. See vol. 2 (Tehran: Farhangsara, 1986), 428.

34. Ibid., 440.

35. Ibid., 435.

36. Sayyid Jamal Turabi Tabataba'i, *Naqshha va Nigashtaha-yi Masjid-i Kabud-i Tabriz* (Tabriz: Shafaq-i Tabriz, 1969), 11. On the history of the Blue Mosque of Tabriz, see Sandra Aube, "La Mosquée bleue de Tabriz (1465): Remarques sur la céramique architecturale Qara Qoyunlu," *Studia Iranica* 37, no. 2 (2008) 241–77; Golombek and Wilber, *Timurid Architecture,* 1: 406–9.

37. Michele Membré, *Mission to the Lord Sophy of Persia, 1539–1542,* trans. A.H. Morton (London: School of Oriental and African Studies, 1993), 51.

38. Aube, "La Mosquée bleue de Tabriz," 249.

39. Charles Melville, "Historical Monuments and Earthquakes in Tabriz," *Iran* 29 (1981): 159–77.

40. Jean Baptiste Tavernier, *Collections of Travels through Turkey into Persia, and the East Indies,* vol. 1 (London: Moses Pitt, 1684), 21.

41. Ibid., 21–22.

42. Tabataba'i, *Naqshha va Nigashtaha-yi Masjid-i Kabud-i Tabriz,* 16.

43. Nadir Mirza, *Tarikh va Jughrafiya-yi Dar al-Saltana-yi Tabriz,* ed. Muhammad Mushiri (Tehran: Iqbal, 1981), 78.

44. Ibid., 79.

45. Eskandar Beg Monshi, *History of Shah 'Abbas the Great,* trans. Roger M. Savory, vol. 2 (Boulder, CO: Westview, 1978), 724. See also Iskandar Bayg Munshi, *Tarikh-i 'Alamara-yi 'Abbasi,* ed. Iraj Afshar, 2 vols. (Tehran: Amir Kabir, 2003), 1:544.

46. On the architecture of Isfahan and its influence on Safavid society and culture, see Babaie, *Isfahan and Its Palaces.*

47. Eskandar Beg Monshi, *History of Shah 'Abbas,* 1038; Iskandar Bayg Munshi, *Tarikh-i 'Alamara-yi 'Abbasi,* 2:831.

48. Eskandar Beg Monshi, *History of Shah 'Abbas,* 1229–30; Iskandar Bayg Munshi, *Tarikh-i 'Alamara-yi 'Abbasi,* 2:1007–8, 1110–11. For a discussion of the construction of Shah 'Abbas's Isfahan and the Shaykh Lutfallah Mosque in particular, see Sheila Canby, *Shah 'Abbas: The Remaking of Iran* (London: British Museum Press, 2009), 22–37.

49. Asher, *Architecture of Mughal India,* 105–10.

50. *The Shah Jahan Nama of 'Inayat Khan: An Abridged History of the Mughal Emperor Shah Jahan, Compiled by His Royal Librarian,* ed. W.E. Begley and Z.A. Desai, trans. A.R. Fuller (Oxford: Oxford University Press, 1990), 211–212. For an account of this monsoon in a Persian text of the Shah Jahan chronicles, see 'Abd al-Hamid Lahori, *Badshah Nama,* ed. Mawlawis Kabir al-Din Ahmad and 'Abd al-Rahim, vol. 1 (Calcutta: College Press, 1867), 276–77.

51. Asher, *Architecture of Mughal India,* 218. On the architectural history of Thatta, see Ahmad Hasan Dani, *Thatta: Islamic Architecture* (Islamabad: Institute of Islamic History, Culture, and Civilization, 1982).

52. See, for instance, his discussions of the Friday Mosque of Bibi Khanum in Samarqand (Kamal al-Din ʿAbd al-Razzaq Samarqandi, *Matlaʿ al-Saʿdayn*, vol. 1, pt. 2, 809) and the Shrine of Khwaja ʿAbdallah Ansari in the Gazargah of Herat (ibid., vol. 2, pt. 2, 861).

4. STONE FROM THE EAST

1. For some recent literature that has reframed and widened the scope of discussions of the age of discovery and exploration to include the empires of Islamic Eurasia, see Palmira Brummett, *Ottoman Seapower and Levantine Diplomacy in the Age of Discovery* (Albany: State University of New York Press, 1994); Muzaffar Alam and Sanjay Subrahmanyam, *Indo-Persian Travels in the Age of Discoveries, 1400–1800* (Cambridge: Cambridge University Press, 2007); Giancarlo Casale, *The Ottoman Age of Exploration* (Oxford: Oxford University Press, 2010).

2. Like porcelain, precious stones were used as ballast to stabilize ships. See Rosamond Mack, *Bazaar to Piazza: Islamic Trade and Italian Art, 1300–1600* (Berkeley: University of California Press, 2002), 250. On the usefulness of jewels and precious stones in separating the bulk of spices in the hold and keeping ships from capsizing, see John Keay, *The Spice Route: A History* (Berkeley: University of California Press, 2006), 227–28.

3. On the connection of the global circulation of objects and commodities to the economic and cultural construction of colors, see Robert Finlay, "Weaving the Rainbow: Visions of Color in World History," *Journal of World History* 18, no. 4 (2007): 383–431; R. Finlay, *The Pilgrim Art: Cultures of Porcelain in World History* (Berkeley: University of California Press, 2010); Sheila S. Blair and Jonathan Bloom, *And Diverse Are Their Hues: Color in Islamic Art and Culture* (New Haven: Yale University Press, 2011); Philip Ball, *Bright Earth: The Invention of Colour* (Chicago: University of Chicago Press, 2001); Victoria Finlay, *Color: A Natural History of the Palette* (New York: Random House, 2002).

4. See Michel Pastoureau, *Blue: The History of a Color* (Princeton: Princeton University Press, 2001). On the question of blue and the history of colors in general, see Ball, *Bright Earth*; V. Finlay, *Color*.

5. R. Finlay, *Pilgrim Art*, 139–40. For other works on natural substances and the making of colors, see Jenny Balfour-Paul, *Indigo* (London: British Museum Press, 1998); Robert Chenciner, *Madder Red, a History of Luxury and Trade* (London: Routledge, 2000).

6. Elisabeth West Fitzhugh and Willem Floor, "Cobalt," *Encyclopaedia Iranica* 5, no. 8 (1992): 873–75.

7. Pastoureau, *Blue*, 26.

8. V. Finlay, *Color*, 280–81.

9. Pastoureau, *Blue*, 21–22.

10. V. Finlay, *Color*, 280–81.

11. Ghirlandaio contract, October 23, 1485, Ospedale degli Innocenti, Florence, quoted in Michael Baxandall, *Painting and Experience in Fifteenth-Century Italy* (Oxford: Oxford University Press, 1972), 6.

12. Ibid., 11.

13. Barbara H. Berrie, "Pigments in Venetian and Islamic Painting," in *Venice and the Islamic World, 828–1797*, ed. Stefano Carboni (New Haven: Yale University Press, 2007), 144.

14. Pastoureau, *Blue*, 72; Berrie, "Pigments in Venetian and Islamic Painting," 145.

15. On the history of indigo, see Balfour-Paul, *Indigo*; Pastoureau, *Blue*; V. Finlay, *Color*.

16. Pastoureau, *Blue*, 125–31.

17. See V. Finlay, *Color*, 339–41.

18. Ruy González de Clavijo, *Embassy to Tamerlane, 1403–1406*, trans. Guy le Strange (London: Routledge, 1928), 181–82.

19. Ibid., 182.

20. Ibid., 270.

21. Ibid., 274.

22. *Travels to Tana and Persia by Josafa Barbaro and Ambrogio Contarini* (1474), ed. Lord Stanley of Alderley, trans. William Thomas (London: Hakluyt Society, 1873), 72–75.

23. Ibid., 56–57, 59.

24. James W. Allan, "Early Safavid Metalwork," in *Hunt for Paradise: Court Arts of Safavid Iran, 1501–1576*, ed. Jon Thompson and Sheila R. Canby (Milan: Skira, 2003), 205.

25. Michele Membré, *Mission to the Lord Sophy of Persia (1539–1542)*, trans. A. H. Morton (London: School of Oriental and African Studies, 1993), 27.

26. See, for instance, Engelbert Kaempfer, *Amoenitatum Exoticarum Politico-Physico-Medicarum Fasciculi V: Variae Relationes, Observationes et Descriptiones Rerum Persicarum et Ulterioris Asiae* (Lemgo, Germany: Heinrich Wilhelm Meyer, 1712), 95.

27. Willem Janszoon Blaeu, *Theatrum Orbis Terrarum*, vol. 2 (Amsterdam: Joan Blaeu, 1635), n.p.

28. Raphaël Du Mans, *Estat de la Perse en 1660*, ed. Charles Schefer (Paris: Ernest Leroux, 1890), 187–88; *Raphaël du Mans: Missionnaire en Perse au XVIIe siècle*, vol. 2, ed. Francis Richard (Paris: L'Harmattan, 1995), 146.

29. *The Travels and Journal of Ambrosio Bembo*, trans. Clara Bargellini (Berkeley: University of California Press, 2007), 336.

30. John Fryer, *A New Account of East India and Persia: Being Nine Years' Travels, 1672–1681* (1698), ed. William Crooke, vol. 3 (London: Hakluyt Society, 1915), 10.

31. Jean Chardin, *Voyages du Chevalier Chardin, en Perse, et Autres Lieux de l'Orient*, 10 vols. (Paris: L. Langlès, 1811), 3:151–52.

32. Ibid., 104.

33. Ibid., 249–54.

34. Ibid., 352–53.

35. Ibid., 353.

36. Ibid., 360–61.

37. Ibid., 7:488.

38. Ibid., 488–89. See also Mas'ud Mirza Zill al-Sultan, *Tarikh-i Sarguzasht-i Mas'udi* (Tehran: Intisharat-i Babak, 1983), 180; Muhammad Ibrahim Bastani Parizi, *Siyasat va Iqtisad-i 'Asr-i Safavi* (Tehran: Intisharat-i Safi 'Ali Shah, 1999), 311.

39. Chardin, *Voyages du Chevalier Chardin*, 7:488.

40. Ibid., 129. See also Mas'ud Mirza Zill al-Sultan, *Tarikh-i Sarguzasht-i Mas'udi*, 180; Bastani Parizi, *Siyasat va Iqtisad-i 'Asr-i Safavi*, 311.

41. Chardin, *Voyages du Chevalier Chardin*, 3:132, 232.

42. Paul E. Desautels, *The Mineral Kingdom* (New York: Madison Square, 1968), 19.

43. See John Mandeville, *The Book of Marvels and Travels*, ed. Anthony Bale (Oxford: Oxford University Press, 2012); Mandeville, *The Travels of Sir John Mandeville*, ed. A.W. Pollard (London: Macmillan, 1915).

44. There are two French editions of the text in the Hoover Collection of Mining and Metallurgy, Honnold Library, Pomona College, Claremont, CA: John Mandeville, *Le Lapidaire en Francoys* (Lyon or Paris: Michel Le Noir, c. 1495–1500) and *Le Lapidaire en Francoys* (Paris: Alain Lotrian, 1544). See also the copy in the Oriental and India Office, British Library, London: Mandeville, *Le Lapidaire en Francoys* (Lyon: Bernardus de Gordonio, c. 1495), IA 4325.

45. Mandeville, *Le Lapidaire en Francoys* (Paris: Alain Lotrian, 1544), fol. C1r–C1v.

46. William Shakespeare, *The Excellent History of the Merchant of Venice* (London: J. Roberts, 1600), act 3, scene 1, lines 105–7.

47. Title page of Camillus Leonardus, *The Mirror of Stones* (London: J. Freeman, 1750), translation of *Speculum Lapidum*.

48. Ibid., 15.

49. Frank J. Anderson, *Riches of the Earth: Ornamental, Precious, and Semiprecious Stones* (New York: Rutledge, 1981), 106–23.

50. Ibid., 118.

51. Desautels, *Mineral Kingdom*, 15.

52. Leonardus, *Mirror of Stones*, 235–36.

53. Camillus Leonardus, *Speculum Lapidum* (Venice: J.B. Sessa, 1502; repr., Paris: Carolus Sevestre, 1610), 417–22.

54. François de la Rue, *De Gemmis Aliquot* (Paris: Chrétien Wechel, 1547), 139–40.

55. Garcia de Orta, *Colloquies on the Simples and Drugs of India*, trans. Sir Clements Markham (London: Henry Sotheran, 1913), 359.

56. Ibid., 358.

57. Ibid, 359.

58. Joseph Pogue, *The Turquois: A Study of Its History, Mineralogy, Geology, Ethnology, Archaeology, Mythology, Folklore and Technology* (Washington DC: National Academy of Sciences, 1915), 17, translation of Boetius de Boot, *Le Parfaict Ioaillier, ou Histoire des Pierreries*, trans. Andre Toll (Lyon: Jean Antoine Huguetan, 1644), 339–40.

59. Pogue, *Turquois*, 17, translation of de Boot, *Le Parfaict Ioaillier*, 346–47.

60. Pogue, *Turquois*, 17, translation of de Boot, *Le Parfaict Ioaillier*, 339.

61. Pogue, *Turquois*, 17, translation of de Boot, *Le Parfaict Ioaillier*, 348–49.

62. Thomas Nichols, *Lapidary; Or, The History of Precious Stones: With Cautions for the Undeceiving of All Those That Deal with Precious Stones* (Cambridge: Thomas Buck, 1652), 147.

63. Ibid., 150–51.

64. Ibid., 148.

65. Ibid., 147.

66. Ibid.

67. Ibid., 146.

68. Ibid., 150.

69. Ibid., 15.

70. Again, it is important to note that marvelous beliefs regarding a magical and astral mineral world persisted and could reach extreme proportions. In 1670, Francesco Lana Terzi, a professor of natural philosophy in Italy, conceived of a far-fetched but wondrous airship, a zeppelin to be carried aloft over the jungles of Brazil by copper spheres and coral-agate powered by the magnetism of the sun: *Podromo, Ouero, Saggio di Alcune Inventioni Nuove Premesso all'Arte Maestra* (Brescia: Rizzardi, 1670), 50–61. Discussed in Anderson, *Riches of the Earth*, 115, 119.

71. Pogue, *Turquois*, 120, translation of de Boot, *Le Parfaict Ioaillier*, 343.

72. Pogue, *Turquois*, 17, translation of de Boot, *Le Parfaict Ioaillier*, 339.

73. Nichols, *Lapidary*, 149.

74. René Antoine de Réaumur, "Observations sur les mines de turquoises du royaume; sur le nature de la matière qu'on y trouve, et sur la manière dont on lui donne la couleur," *Memoires de l'académie royale des sciences* (1715): 174–202.

75. Jean-Antoine Chaptal, *Elements of Chemistry*, trans. William Nicholson (Boston: J.T. Buckingham, 1806; originally published as *Éléments de Chimie* [Montpellier: Jean-François Picot, 1790]), 360.

76. [Edme-Jean Baptiste] Bouillon-Lagrange, "Analyse d'une substance connue sous le nom de *turquoise*," *Annales de Chimie, ou Recueil de Mémoires Concernant la Chimie et les Arts Qui en Dépendent* 59 (1806): 188.

77. Georgius Agricola, *De Re Metallica* (Basel: Froben, 1556); Desautels, *Mineral Kingdom*, 18.

78. Desautels, *Mineral Kingdom*, 18–20.

79. James Sowerby, *Exotic Mineralogy: Or, Coloured Figures of Foreign Minerals as a Supplement to British Mineralogy*, vol. 1 (London: Benjamin Meredith, 1811), 183.

80. Pogue, *Turquois*, 27.

81. Isaac Newton, *Opticks: Or, A Treatise of the Reflexions, Refractions, Inflexions and Colours of Light* (London, 1704), cited in V. Finlay, *Color*, 6–8.

82. Desautels, *Mineral Kingdom*, 108–9.

83. Johann Wolfgang von Goethe, *Goethe's Theory of Colours*, trans. Charles Lock Eastlake (London: John Murray, 1840); Pastoureau, *Blue*, 137–38.

84. For Haüy's models of the structure of crystals, see his *Traité de Minéralogie*, 2nd ed. (Paris: Delance, 1822), vol. 5, plate 2. See also Desautels, *Mineral Kingdom*, 36–38.

85. Abbé Haüy, *Traité des Caractères Physiques de Pierres Précieuses* (Paris: Courcier, 1817), 62–63. See also Haüy, *Traité de Minéralogie*, 3:405, 516–17.

5. THE OTHER SIDE OF THE WORLD

1. On the economy and trade of nineteenth-century Iran and its integration into the world economy, see Charles Issawi, *The Economic History of Iran, 1800–1914* (Chicago: University of Chicago Press, 1971); Abbas Amanat, *Cities and Trade: Consul Abbott on the Economy and Society of Iran, 1847–1866* (London: Ithaca, 1983); Gad Gilbar, "The Opening Up of Qajar Iran: Some Economic and Social Aspects," *Bulletin of the School of Oriental and African Studies* 49, no. 1 (1986): 76–89; Peter Avery, Gavin Hambly, and Charles Melville, eds., *From Nadir Shah to the Islamic Republic*, vol. 7 of *The Cambridge History of Iran* (Cambridge: Cambridge University Press, 1991).

2. On the nineteenth-century globalization of the networks of Islam, see Nile Green, *Bombay Islam: The Religious Economy of the West Indian Ocean, 1840–1915* (Cambridge: Cambridge University Press, 2011); Green and James Gelvin, eds., *Global Muslims in the Age of Steam and Print* (Berkeley: University of California Press, 2014).

3. John Mawe, *A Treatise on Diamonds and Precious Stones, Including Their History—Natural and Commercial* (London: Longman, Hurst, Rees, Orme, and Brown, 1813), 153–54.

4. Ibid.

5. Ibid., v–vi, 154.

6. For instance, see Gotthelf Fischer, *Essai sur le turquoise* (Moscow: Imprimerie de l'Université Impériale, 1816).

7. Elise Bee, "The Gems of the Christmas Month," *Jewellers Circular and Horological Review* 29, no. 18 (November 28, 1894): 17.

8. Harry Emanuel, *Diamonds and Precious Stones: Their History, Value, and Distinguishing Characteristics, with Simple Tests for Their Identification* (London: John Camden Hotten, 1867), iii.

9. Ibid., 181.

10. Ibid.; Edward Balfour, ed., *Cyclopaedia of India and of Eastern and Southern Asia*, vol. 5 (Madras: Lawrence and Adelphi, 1873), 234.

11. Mohammed Ben Manssur, "Extracts from the Persian Work Called 'The Book of Precious Stones,' by Mohammed Ben Manssur," *Annals of Philosophy; or, Magazine of Chemistry, Mineralogy, Mechanics, Natural History, Agriculture and the Arts* 15 (1820): 178–90.

12. Emanuel, *Diamonds and Precious Stones*, 22, 181.

13. Charles Hetherington, *Selim, the Nasakchi, a Persian Tale, in Verse* (London: Whittaker, 1867), 31.

14. For the volumes focusing on natural history, see *Description de l'Égypte ou Recuil des Observations et des Recherches Qui Ont Été Faites en Égypte Pendant l'Expedition de l'Armée Française: Histoire Naturelle*, 2 vols. (Paris: Imprimerie Impériale, 1813). For analysis, see Edward Said, *Orientalism* (New York: Vintage, 1978), 42, 84–87; Timothy Mitchell, *Colonising Egypt* (Berkeley: University of California Press, 1991), 31, 46.

15. On rivers, empire, and the environmental history of Ottoman Egypt, see Alan Mikhail, *Nature and Empire in Ottoman Egypt* (Cambridge: Cambridge University Press, 2011).

16. For some accounts of mining in ancient Egypt, see James Henry Breasted, *A History of Egypt, from the Earliest Times to the Persian Conquest* (New York: Charles Scribner's Sons, 1905); Lina Eckenstein, *A History of Sinai* (London: Macmillan, 1921); Sir Alan Gardiner, *Egypt of the Pharaohs: An Introduction* (Oxford: Oxford University Press, 1961); Nicolas Grimal, *A History of Ancient Egypt* (Oxford: Oxford University Press, 1988); Lise Manniche, *Sacred Luxuries: Fragrance, Aromatherapy, and Cosmetics in Ancient Egypt* (Ithaca: Cornell University Press, 1999); Ian Shaw, ed., *The Oxford History of Ancient Egypt* (Oxford: Oxford University Press, 2000).

17. Edward Henry Palmer, *The Desert of the Exodus: Journeys on Foot in the Wilderness of the Forty Years' Wanderings; Undertaken in Connexion with the Ordnance Survey of Sinai and the Palestine Exploration Fund*, vol. 2 (Cambridge: Deighton, Bell, 1871), 232.

18. Jaromir Malek, "The Old Kingdom (*c*.2686–2160 BC)," in Shaw, *Oxford History of Ancient Egypt*, 105.

19. Gae Callender, "The Middle Kingdom Renaissance (*c*.2055–1650 BC)," in Shaw, *Oxford History of Ancient Egypt*, 168.

20. Charles W. Wilson, *Picturesque Palestine: Sinai and Egypt*, vol. 4 (London: J. S. Virtue, 1881), 52.

21. Gaston Camille Charles Maspero, *History of Egypt, Chaldea, Syria, Babylonia, and Assyria*, trans. M. L. McClure, vol. 2 (London: Grolier Society, 1903), 333.

22. Ibid., 333–35; Eckenstein, *History of Sinai*, 23.

23. W. M. Flinders Petrie, *Researches in Sinai* (London: E. P. Dutton, 1906), 108; Joseph Pogue, *The Turquois: A Study of Its History, Mineralogy, Geology, Ethnology, Archaeology, Mythology, Folklore and Technology* (Washington DC: National Academy of Sciences, 1915), 31.

24. Petrie, *Researches in Sinai*, 152–53, 187, 191.

25. Léon de Laborde and Louis Maurice Adolphe Linant de Bellefonds, *Voyage de l'Arabie Pétrée* (Paris: Giard, 1830), 84; Carl Ritter, *The Comparative Geography of Palestine and the Sinaitic Peninsula*, vol. 1 (London: J. B. Lippincott, 1865), 355; Pogue, *Turquois*, 33.

26. Laborde and Linant, *Voyage de l'Arabie Pétrée*, 84; Manniche, *Sacred Luxuries*, 45.

27. Laborde and Linant, *Voyage de l'Arabie Pétrée*, 84.

28. Wilson, *Picturesque Palestine*, 13.

29. Palmer, *Desert of the Exodus*, 230.

30. Wilson, *Picturesque Palestine*, 32.

31. Maspero, *History of Egypt*, 166–67.

32. Wilson, *Picturesque Palestine*, 34–35.

33. Palmer, *Desert of the Exodus*, 231.

34. Wilson, *Picturesque Palestine*, 52.

35. Palmer, *Desert of the Exodus*, 202.

36. Maspero, *History of Egypt*, 161–62.

37. Petrie, *Researches in Sinai,* 48–50.

38. Maspero, *History of Egypt,* 165–66.

39. Pogue, *Turquois,* 32.

40. Maspero, *History of Egypt,* 166.

41. Petrie, *Researches in Sinai,* 50–51.

42. Heinrich Brugsch, *Wanderung nach den Turkis-Minen und der Sinai-Halbinsel* (Leipzig: J.T. Hinrichs, 1866). The account presented here draws on John Cooney's "Major Macdonald, a Victorian Romantic," *Journal of Egyptian Archaeology* 58 (August 1972): 280–85, based largely on Brugsch's text.

43. Cooney, "Major Macdonald," 281.

44. Charles Macdonald, quoted in Robert Hunt, *Ure's Dictionary of Arts, Manufactures, and Mines, Containing a Clear Exposition of Their Principles and Practice* (London: Longmans, Green, 1878), 1048.

45. L. Feuchtwanger, *A Popular Treatise on Gems in Reference to Their Scientific Values* (New York: A. Hanford, 1859), 332.

46. Cooney, "Major Macdonald," 283.

47. *Exhibition of the Works of Industry of All Nations in 1851: Reports by the Jurists on the Subjects in the Thirty Classes into Which the Exhibition Was Divided* (London: William Clowes and Sons, 1852), 7.

48. Ibid.

49. "The Peninsula of Sinai; Notes of Travel Therein," *Littell's Living Age* 1133 (February 17, 1866): 489.

50. Wilson, *Picturesque Palestine,* 50.

51. "Peninsula of Sinai," 490.

52. Cooney, "Major Macdonald," 282.

53. Emanuel, *Diamonds and Precious Stones,* 180.

54. Palmer, *Desert of the Exodus,* 201.

55. George Ebers, *Durch Gosen zum Sinai* (Leipzig: W. Engelmann, 1872), 136; Cooney, "Major Macdonald," 282–83.

56. Petrie, *Researches in Sinai,* 53.

57. Petrie, *Researches in Sinai,* 53.

58. Palmer, *Desert of the Exodus,* 202.

59. Ibid., 201.

60. Wilson, *Picturesque Palestine,* 52.

61. Palmer, *Desert of the Exodus,* 197.

62. Petrie, *Researches in Sinai,* 46.

63. Ibid., 48.

64. William Ouseley, *Travels in Various Countries of the East; More Particularly Persia,* vol. 1 (London: Rodwell and Martin, 1819), 210.

65. Ibid., 210–11.

66. Pascal Coste, *Monuments Modernes de la Perse: Mesurés, Dessinés et Décrits* (Paris: A. Morel, 1867). See also Eugène Flandin and Coste, *Voyage en Perse: Perse Moderne,* vol. 8 (Paris: Gide and J. Baudry, 1851).

67. Review of *Narrative of a Journey into Khorasan, in the Years 1821 and 1822,* in *Asiatic Journal and Monthly Miscellany, East India Company,* vol. 20 (London, 1825), 555.

68. James Baillie Fraser, *Narrative of a Journey into Khorasan, in the Years 1821 and 1822; Including Some Account of the Countries to the North-East of Persia* (London: Longman, Hurst, Rees, Orme, Brown, and Green, 1825), 409.

69. James Baillie Fraser, *Travels and Adventures in the Persian Provinces on the Southern Banks of the Caspian Sea* (London: Longman, Rees, Orme, Brown, and Green, 1826), 344.

70. On the Badakhshan migrants, see J.P. Ferrier, *Caravan Journeys and Wanderings in Persia, Afghanistan, Turkistan, and Beloochistan*, trans. William Jesse (London: John Murray, 1856), 106, quoting Alexandre Chodsko.

71. Ibid., 107.

72. Fraser, *Narrative*, 410.

73. Ibid., 411; Fraser, *Travels and Adventure*, 345.

74. Fraser, *Narrative*, 412; Fraser, *Travels and Adventure*, 345.

75. Fraser, *Narrative*, 412; Fraser, *Travels and Adventure*, 345.

76. Fraser, *Narrative*, 414; Fraser, *Travels and Adventure*, 346.

77. Fraser, *Narrative*, 413; Fraser, *Travels and Adventure*, 346.

78. Fraser, *Travels and Adventure*, 343.

79. Fraser, *Narrative*, 416.

80. Ibid., 410.

81. Ibid., 416; Fraser, *Travels and Adventure*, 346–47.

82. Chodsko, quoted in Ferrier, *Caravan Journeys*, 106.

83. Mohan Lal, *Travels in the Panjab, Afghanistan, and Turkistan, to Balk, Bokhara, and Herat* (London: Wm.H. Allen, 1846), 175.

84. Ibid., 177.

85. Ibid., 176.

86. Fraser, *Narrative*, 416.

87. Ibid., 417.

88. Ibid.

89. Ibid., 418–19.

90. Ibid., 468.

91. Ibid., 469.

92. Ibid.

93. Ibid.

94. Ibid., 419.

95. *Tarikh al-Hind al-Gharbi al-Musamma bi-Hadith-i Naw* (Istanbul: Ibrahim Muteferrika, 1729).

96. For examples of this literature, see Mirza Abu al-Hasan Khan Ilchi, *Hayratnama: Safarnama-yi Mirza Abu al-Hasan Khan Ilchi bih Landan*, ed. Hasan Mursilvand (Tehran: Farhang-i Rasa, 1985); Mirza I'tisam al-Din, *Shigurf Namah i Velaet, or Excellent Intelligence Concerning Europe; Being the Travels of Mirza Itesa Modeen, in Great Britain and France*, trans. James Edward Alexander (London: Parbury, Allen, 1827). For analysis of Qajar travel literature about Europe, see Mohamad Tavakoli-Targhi, *Refashioning Iran: Orientalism, Occidentalism, and Historiography* (Basingstoke: Palgrave Macmillan, 2001); Naghmeh Sohrabi, *Taken for Wonder: Nineteenth-Century Travel Accounts from Iran to Europe* (Oxford: Oxford University Press, 2012).

97. Muhammad Hasan Khan Sani'al-Dawla I'timad al-Saltana, "Tarikh-i Inkishaf-i Yingi Dunya" (A.H. 1288/1871), MS 1138797, National Library and Archives of Iran (Sazman-i Asnad va Kitabkhana-yi Milli), Tehran.

98. Ibid., fols. 86, 88.

99. Ibid., fols. 94–105.

100. Ibid., fols. 108–20.

101. Ibid., fols. 120–22.

102. On the Qajar conception of "the guarded domains of Iran" (*mamalik-i mahrusa-yi Iran*), see Abbas Amanat, *Pivot of the Universe: Nasir al-Din Shah and the Iranian Monarchy, 1831–1896* (Berkeley: University of California Press, 1997), 13–18.

103. *Ruznama-yi Vaqa'i'-yi Ittifaqiya* 44 (A.H. 1268/1851): 5, reprinted in a collection of the same name, 4 vols. (Tehran: National Library of the Islamic Republic of Iran, 1994), 1:233. (Subsequent notes give the reprint volume and page numbers in parentheses.) This gazetteer later changed its name to *Ruznama-yi Dawlat-i 'Awliya-yi Iran*.

104. *Ruznama-yi Vaqa'i'-yi Ittifaqiya* 83 (A.H. 1268/1851), 5 (1:492).

105. Muhammad Hasan Khan Sani'al-Dawla I'timad al-Saltana, *Tarikh-i Muntazam-i Nasiri,* ed. Muhammad Isma'il Rizvani, vol. 3 (Tehran: Dunya-yi Kitab, 1963), 1753.

106. *Ruznama-yi Vaqa'i'-yi Ittifaqiya* 152 (A.H. 1270/1853), 6 (2:963).

107. I'timad al-Saltana, *Tarikh-i Muntazam-i Nasiri,* 1704.

108. Ibid., 1711.

109. *Ruznama-yi Vaqa'i'-yi Ittifaqiya* 74 (A.H. 1268/1851), 5 (1:435).

110. *Ruznama-yi Vaqa'i'-yi Ittifaqiya* 58 (A.H. 1268/1851), 5 (1:317).

111. *Ruznama-yi Vaqa'i'-yi Ittifaqiya* 118 (A.H. 1269/1852), 7 (1:723). The muhur (from *muhr*, Persian for "seal") was equal to fifteen silver rupees.

112. *Ruznama-yi Vaqa'i'-yi Ittifaqiya* 127 (A.H. 1269/1852), 8 (1:788). See also *Ruznama-yi Vaqa'i'-yi Ittifaqiya* 134 (A.H. 1269/1852), 3 (2:835).

113. The literature on Western imperial surveying, knowledge production, and collection of information is extensive. For some examples, see Said, *Orientalism;* Said, *Culture and Imperialism* (New York: Vintage, 1994); Mary Louise Pratt, *Imperial Eyes: Travel Writing and Transculturation* (London: Routledge, 1992); Thomas Metcalf, *Ideologies of the Raj* (Berkeley: University of California Press, 1995); C. A. Bayly, *Empire and Information* (Cambridge: Cambridge University Press, 1996).

114. On the life of I'timad al-Saltana, his relations with the shah, and his geographical projects, see Amanat, *Pivot of the Universe,* 62, 432; Firoozeh Kashani-Sabet, *Frontier Fictions: Shaping the Iranian Nation, 1804–1946* (Princeton: Princeton University Press, 1999), 41–43, 64–65.

115. Muhammad Hasan Khan Sani'al-Dawla I'timad al-Saltana, *Matla' al-Shams, Tarikh-i Arz-i Aqdas va Mashhad-i Muqaddas, dar Tarikh va Jughrafiya-yi Mashruh-i Balad va Imakan-i Khurasan* (1882–84), 3 vols. (Tehran: Farhangsara, 1986). I'timad al-Saltana also recorded the shah's oral travel narrative, the text lithographed in nastaliq script by Mirza Riza Kalhur. See Nasir al-Din Shah Qajar, *Safarnama-yi Duvvum-i Khurasan* (1882; Tehran: Intisharat-i Kavus, 1984).

116. For other recent studies that similarly suggest that the literature of travel and exploration could indeed reflect the world it set out to observe and was not merely a dead-end journey into the mind-set and culture of the observer, see Harold J. Cook, *Matters of Exchange: Commerce, Medicine, and Science in the Dutch Golden Age* (New Haven: Yale University Press, 2007), 5–6, 21; Rudolph P. Matthee, "The Safavids under Western Eyes: Seventeenth-Century European Travelers to Iran," *Journal of Early Modern History* 13 (2009): 137–71.

117. Albert Houtum-Schindler, "The Turquoise Mines of Nishapur, Khorassan," in *Records of the Geological Survey of India*, vol. 17, 132–42 (Calcutta: Geological Survey, 1884). Houtum-Schindler's travelogue on Khurasan also mentions these mines: see "Safarnama-yi Khurasan," in *Sih Safarnama*, ed. Qudrat Allah Rushani Za'faranlu (Tehran: University of Tehran, 1968), 145–211.

118. See Edward G. Browne, "The Persian Manuscripts of the Late Sir Albert Houtum-Schindler," *Journal of the Royal Asiatic Society of Great Britain and Ireland* 49 (October 1917): 657–94. On Houtum-Schindler's life and career, see John Gurney, "Albert Houtum-Schindler," *Encyclopaedia Iranica* 12, no. 5 (2003): 540–43.

119. I'timad al-Saltana, *Matla' al-Shams*, 3:863.

120. Ibid.

121. Ibid., 864.

122. Ibid., 864–65.

123. Jean Chardin, *Voyages du chevalier Chardin, en Perse, et Autres Lieux de l'Orient*, vol. 3 (Paris: L. Langlès, 1811), 360n.

124. I'timad al-Saltana, *Matla' al-Shams*, 3:865.

125. Ibid., 866.

126. Ibid., 866–67.

127. Ibid., 867.

128. Ibid., 868.

129. Ibid., 878.

130. Ibid., 864.

131. Ibid.

132. Mirza Husayn bin 'Abd al-Karim Durrudi, *Kitabcha-yi Nishabur: Guzarish-i Rustha-yi Nishapur dar Sal-i 1296 Qamari* (1878), ed. Rasul Ja'fariyan (Mashhad: Astan-i Quds-i Razavi, 2003), 140–48.

133. I'timad al-Saltana, *Matla 'al-Shams*, 3:869–70.

134. Ibid., 867.

135. 'Abdallah Khan Qajar, Private Photographer of His Majesty the Shah, *Mining for Turquoise Stones*, photograph in album 291 (A.H. 1312/1894), 72, Gulistan Palace Museum, Tehran.

136. Ibid., 72.

137. Ibid.

138. I'timad al-Saltana, *Matla' al-Shams*, 3:869–70.

139. Ibid., 867.

140. Ibid., 870.

141. Ibid.; Houtum-Schindler, "The Turquoise Mines of Nishapur," 138. 'Abdallah Khan Qajar estimated the daily earnings of the miners as three thousand dinars, the equivalent of three qiran. See Gulistan album 291, 72.

142. Houtum-Schindler, "Turquoise Mines of Nishapur," 134, 137.

143. I'timad al-Saltana, *Matla' al-Shams*, 3:878.

144. Ibid., 870–72.

145. Ibid., 871.

146. For the account of a late nineteenth-century jewel merchant's journey from Tehran to Europe, curiously lacking in mineralogical detail on gems and their value and trade, see Ibrahim Sahhafbashi, *Safarnama-yi Ibrahim Sahhafbashi*, ed. Muhammad Mushiri (Tehran: Shirkat-i Mu'allifan va Mutarjiman-i Iran, 1978). For analyses of Sahhafbashi's travels and encounters in Europe, see Tavakoli-Targhi, *Refashioning Iran*, 62, 71; Sohrabi, *Taken for Wonder*, 105–24.

147. I'timad al-Saltana, *Matla' al-Shams*, 3:871; Houtum-Schindler, "Safarnama-yi Khurasan," 178.

148. I'timad al-Saltana, *Matla' al-Shams*, 3:871.

149. *Ruznama-yi Dawlat-i 'Awliya-yi Iran* 356 (A.H. 1274/1857): 1, reprinted in a collection of the same name, 2 vols. (Tehran: National Library of the Islamic Republic of Iran, 1991), 2:XXX. (Subsequent notes give the reprint volume and page numbers in parentheses.)

150. *Ruznama-yi Dawlat-i 'Awliya-yi Iran* 537 (A.H. 1279/1863): 2 (2:XXX).

151. For instance, see I'timad al-Saltana, *Matla' al-Shams*, 3:864, 879.

152. Jean-Baptiste Feuvrier, *Trois ans à la cour de Perse* (Paris: Imprimerie Nationale, 1900), 132; Houtum-Schindler, "Turquoise Mines of Nishapur," 137. I'timad al-Saltana, *Matla' al-Shams*, 3:863–79, lists the annual rent of the turquoise mines as eight hundred to twelve hundred tumans.

153. Houtum-Schindler, "Turquoise Mines of Nishapur," 137, 140.

154. Ibid., 141. See also Gurney, "Albert Houtum-Schindler," 540–41.

155. See, for instance, Browne, "Persian Manuscripts," 657–62; George Nathaniel Curzon, *Persia and the Persian Question*, vol. 1 (London: Longmans, Green, 1889), xiii, 477.

156. Gurney, "Albert Houtum-Schindler," 542.

157. In return, however, their yearly rate of taxation was lowered from eight thousand to five thousand tumans. Houtum-Schindler, "Turquoise Mines of Nishapur," 141.

158. Ibid.

159. Ibid.

160. Ibid.; I'timad al-Saltana, *Matla' al-Shams*, 3:871.

161. Curzon, *Persia and the Persian Question*, 266.

162. Ibid., 264.

163. E.C. Ringler Thomson, "Report on the Trade and Commerce of Khorasan for the Financial Year 1895–96," Mashhad, June 1, 1896, 25, Foreign Office—Diplomatic and Consular Reports, Annual Reports, 1800, British National Archives (hereafter BNA), Kew Gardens. For a more personal and sardonic version of the vice-consul's report on the turquoise mines, see Ringler Thomson, "The Turquoise Mines of Persia," *Windsor Magazine* 26 (1907): 61–67.

164. E.C. Ringler Thomson to H.B.M. envoy extraordinary and minister plenipotentiary at the court of Persia in Tehran, April 14, 1895, Mashhad Extracts News Letters, April 20, 1895, FO 248/612 (1895), BNA.

165. Ringler Thomson, "Report on the Trade and Commerce of Khorasan," 25.

166. Ibid., 28–29.

167. Ibid., 28.

168. Bee, "Gems of the Christmas Month," 17.

169. Ringler Thomson, "Report on the Trade and Commerce of Khorasan," 29.

170. Ringler Thomson, "Turquoise Mines of Persia," 65.

171. Ringler Thomson, "Report on the Trade and Commerce of Khorasan," 28.

EPILOGUE

1. Bernardino de Sahagún, "*Book Twelve of the Florentine Codex,*" in *We People Here: Nahuatl Accounts of the Conquest of Mexico,* trans. and ed. James Lockhart (Berkeley: University of California Press, 1993), 65.

2. Ibid., 252–53.

3. Ibid., 248.

4. Colin McEwan, Andrew Middleton, Caroline Cartwright, and Rebecca Stacey, *Turquoise Mosaics from Mexico* (Durham: Duke University Press, 2006), 27.

5. Ibid., 8.

6. Ibid. 27–30.

7. This southward flow remains the conventional view even as scholars search for hints of stones mined closer to the Aztecs. On the American turquoise trade in the time of the Aztecs, see Phil Weigand and Acelia Garcia de Weigand, "A Macroeconomic Study of the Relationship between the Ancient Cultures of the American Southwest and Mesoamerica," in *The Roads to Aztlan: Art from a Mythic Homeland,* eds. Virginia Fields and Victor Zamudio-Taylor (Los Angeles: Los Angeles County Museum of Art, 2001), 184–95.

8. McEwan et al., *Turquoise Mosaics from Mexico,* 54–59.

9. Ibid., 10.

10. For a modern edition of the text, see *The Codex Mendoza,* eds. Frances Berdan and Patricia Rieff Anawalt, 4 vols. (Berkeley: University of California Press, 1992). For a discussion of turquoise in the Codex Mendoza, see McEwan et al., *Turquoise Mosaics from Mexico,* 10, 27.

11. *Codex Mendoza,* 2:20, 25, 113; McEwan et al., *Turquoise Mosaics from Mexico,* 21.

12. Bernardino de Sahagún, *General History of the Things of New Spain,* trans. Arthur J. O. Anderson and Charles Dibble, 2nd ed., 12 vols. (Santa Fe: School of American Research, 1975), vol. 1, plate 2, fig. 2. See also McEwan et al., *Turquoise Mosaics from Mexico,* 59.

13. Sahagún, *General History,* 11:221.

14. Ibid., 221–22.

15. Ibid., 222–24.

16. B. Silliman, "Turquoise of New Mexico," *American Journal of Science and Arts* 122 (1881): 69.

17. W.P. Blake, "The Chalchihuitl of the Ancient Mexicans: Its Locality and Association, and Its Identity with Turquoise," *American Journal of Science and Arts* 25 (1858): 227.

18. Ibid., 230.

19. Ibid., 227.

20. F.H. Cushing, "Outlines of Zuni Creation Myths," in *Thirteenth Annual Report of the Bureau of Ethnology, for the Years 1891–92*, ed. J.W. Powell (Washington DC: Smithsonian Institution, 1896), 369; Joseph Pogue, *The Turquois: A Study of Its History, Mineralogy, Geology, Ethnology, Archaeology, Mythology, Folklore and Technology* (Washington DC: National Academy of Sciences, 1915), 123; F.W. Hodge, *Turquois Work of the Hawikuh, New Mexico* (New York: Museum of the American Indian, 1921), 7.

21. Hodge, *Turquois Work of the Hawikuh*, 5.

22. Ibid., 7.

23. Pogue, *Turquois*, 125; Washington Matthews, *Navaho Legends* (New York: Houghton, Mifflin, 1897), 163.

24. Blake, "Chalchihuitl of the Ancient Mexicans," 228.

25. Ibid., 227.

26. Pogue, *Turquois*, 53–55.

27. Ibid., 135.

28. Fayette Alexander Jones, *New Mexico Mines and Minerals* (Santa Fe: New Mexican Printing Company, 1904), 274.

29. A. Carnot, "Sur la composition chimique des turquoises," *Bulletin de la Société française de minéralogie et de cristallographie* 18 (1895): 119–23.

30. Jones, *New Mexico Mines and Minerals*, 271.

31. Department of the Interior, United States Geological Survey, *Mineral Resources of the United States* (Washington DC: Government Printing Office) for 1888, 582; 1889–90, 446; 1892, 763–63. See also Pogue, *Turquois*, 135.

Bibliography

ARCHIVES AND COLLECTIONS
British National Archives, Kew Gardens
Foreign Office—Diplomatic and Consular Reports, Annual Series
Ringler Thomson, E. C. "Report on the Trade and Commerce of Khorasan for the Financial Year 1895–96." Mashhad, June 1, 1896. FO Diplomatic and Consular Reports, Annual Reports, 1800.

Foreign Office 248—Meshed Correspondence
FO 248/503 (1890); FO 248/504 (1890); FO 248/529 (1891); FO 248/545 (1892); FO 248/546 (1892); FO 248/568 (1893); FO 248/569 (1893); FO 248/570 (1893); FO 248/592 (1894); FO 248/611 (1895); FO 248/612 (1895); FO 248/613 (1895); FO 248/632 (1896); FO 248/652 (1897); FO 248/653 (1897); FO 248/674 (1898); FO 248/696 (1899); FO 248/697 (1899).

Cambridge University Library
Edward G. Browne Collection—Persian Manuscripts
Anonymous. "Majmu 'at al-Sana 'i" (A.H. 1259/1843). MS Browne P. 32 (9).
Muhammad ibn al-Mubarak al-Qazvini. "Risala dar Ma 'rifat-i Javahir" (A.H. 1300/1883). MS Browne P. 29 (9).
Muhammad ibn Mansur. "Javahirnama" (A.H. 1259/1843). MS Browne P. 32 (9).
———. "Javahirnama" (A.H. 1260/1844). MS Browne P. 31 (9).
———. "Javahirnama" (A.H. 1300/1883). MS Browne P. 29 (9).

Nasir al-Din Tusi. "Tansuqnama-yi Ilkhani" (A.H. 973/1566). MS Browne P. 30 (8).
———. "Tansuqnama-yi Ilkhani" (A.H. 1300/1883). MS Browne P. 29 (9).
Zayn al-Din Muhammad Jami. "Mukhtasar dar bayan-i shinakhtan-i javahir" (A.H. 1259/1843). MS Browne P. 32 (9).

Gulistan Palace Museum, Tehran
Album 291 (A.H. 1312/1894), 72.
Album 296 (A.H. 1300/1883), 14.
Album 296 (A.H. 1300/1883), 64.
Album 374 (A.H. 1279/1863), 18.

Majlis Library Archives (Kitabkhana-yi Majlis), Tehran
Anonymous. "Risala dar Ma'rifat-i Filizha va al-Hajjar" (n.d.). MS 8803.
'Attar bin Muhammad. "Kitab al-Khass al-Ahjar" (A.H. 1176/1861). MS 11700.
Muhammad Ibn Mansur. "Haqidat al-Javahir" (n.d.). MS 5684.
———. "Javahirnama" (n.d.). MS 9103.
———. "Javahirnama" (A.H. 1230/1815). MS 2166.
———. "Javahirnama" (n.d.) MS 15062.
———. "Kitab dar Javahirat va Filizat va Khass-i Har Yak" (A.H. 1283/1868). MS 2167.
———. "Risala-yi Iskandariya va Javahirnama" (n.d.). MS 5690/3.

National Library and Archives of Iran (Sazman-i Asnad va Kitabkhana-yi Milli), Tehran
I'timad al-Saltana, Muhammad Hasan Khan Sani' al-Dawla. "Tarikh-i Inkishaf-i Yingi Dunya" (A.H. 1288/1871). MS 1138797.

Oriental and India Office, British Library, London
Muhammad Ibn Ashraf Rustamdari. "Nuskha-yi Javahirnama-yi Humayuni" (A.H. 1268/1852). MS Or. 1717.

PRINTED TEXTS AND EDITIONS
'Abd al-Hamid Lahori, *Badshah Nama*. Edited by Mawlawis Kabir al-Din Ahmad and 'Abd al-Rahim. 3 vols. Calcutta: College Press, 1867.
'Abd al-Karim Durrudi, Mirza Husayn bin. *Kitabcha-yi Nishabur: Guzarish-i Rustha-yi Nishapur dar Sal-i 1296 Qamari* (1878). Edited by Rasul Ja'fariyan. Mashhad: Astan-i Quds-i Razavi, 2003.
Abolqasem Ferdowsi. *Shahnameh: The Persian Book of Kings*. Translated by Dick Davis. New York: Penguin, 2006.

Abu-l-Fazl. *The Akbar Nama of Abu-l-Fazl: History of the Reign of Akbar Including an Account of His Predecessors.* Translated by H. Beveridge. 2 vols. Calcutta: Asiatic Society of Bengal, 1902.

Abu'l Qasim 'Abdallah Kashani. *'Ara'is al-Javahir va Nafa'is al-Ata'ib.* Edited by Iraj Afshar. Tehran: Bahman, 1966.

Abu Mansur 'Abd al-Malik al-Tha'alibi. *The Lata'if al-Ma'arif of Tha'alibi: The Book of Curious and Entertaining Information.* Translated by C.E. Bosworth. Edinburgh: Edinburgh University Press, 1968.

Abu Rayhan al-Biruni. *Kitab al-Jamahir fi Ma'rifat al-Jawahir.* Translated by Hakim Muhammad Said. Islamabad: Pakistan Hijra Council, 1989.

Agricola, Georgius. *De Re Metallica.* Basel: Froben, 1556.

'Ala' ibn al-Husayn al-Bayhaqi. *Ma'din al-Nawadir fi Ma'rifat al-Jawahir.* Kuwait: Dar al-'Arabiya li al-Nashr wa'l-Tawzi', 1985.

Albuquerque, Afonso de. *The Commentaries of the Great Afonso Dalboquerque, Second Viceroy of India.* Translated by Walter De Gray Birch. 4 vols. London: Hakluyt Society, 1875.

Amin Ahmad Razi. *Haft Iqlim.* Edited by Javad Fazil. 3 vols. Tehran: 'Ali Akbar 'Ilmi, n.d.

Amin Sayyid 'Abdallah al-Husayni Ma'ruf bi Asil al-Din Va'iz Haravi. "Maqsad al-Iqbal-i Sultaniyya." In *Risala-yi Mazarat-i Herat,* edited by Fikri Saljuqi, vol. 1, pt. 1, 1–103. Kabul: Publishing Institute, 1967.

Arakel of Tabriz: The History of Vardapet Arakel of Tabriz. Translated by George Bournoutian. 2 vols. Costa Mesa, CA: Mazda, 2006.

Asiatic Journal and Monthly Miscellany, East India Company. Vol. 20. London, 1825.

Balfour, Edward, ed. *Cyclopaedia of India and of Eastern and Southern Asia.* Vol. 5. Madras: Lawrence and Adelphi, 1873.

Barbaro, Josafa, and Ambrogio Contarini. *Travels to Tana and Persia by Josafa Barbaro and Ambrogio Contarini.* Edited by Lord Stanley of Alderley. Translated by William Thomas. London: Hakluyt Society, 1873.

Bee, Elise. "The Gems of the Christmas Month." *Jewellers Circular and Horological Review* 29, no. 18 (November 28, 1894): 17–19.

Bembo, Ambrosio. *The Travels and Journal of Ambrosio Bembo.* Translated by Clara Bargellini. Berkeley: University of California Press, 2007.

Blaeu, Willem Janszoon. *Theatrum Orbis Terrarum.* Vol. 2. Amsterdam: Joan Blaeu, 1635.

Blake, W.P. "The Chalchihuitl of the Ancient Mexicans: Its Locality and Association, and Its Identity with Turquoise." *American Journal of Science and Arts* 25 (1858): 227–32.

Bouillon-Lagrange, [Edme-Jean Baptiste]. "Analyse d'une substance connue sous le nom de *turquoise.*" *Annales de Chimie, ou Recueil de Mémoires Concernant la Chimie et les Arts Qui en Dépendent* 59 (1806): 180–201.

Brugsch, Heinrich. *Wanderung nach den Turkis-Minen und der Sinai-Halbinsel.* Leipzig: J.T. Hinrichs, 1866.

Budaq Munshi Qazvini. *Kitab-i Javahir al-Akhbar.* Edited by Muhsin Bahram Nijhad. Tehran: Ayina-yi Miras, 1999.

Campbell, A. "Notes on Eastern Tibet." *Journal of the Asiatic Society of Bengal* 21 (1855): 215–328.

Carnot, A. "Sur la composition chimique des turquoises." *Bulletin de la Société française de minéralogie et de cristallographie* 18 (1895): 119–23.

Chaptal, Jean-Antoine. *Elements of Chemistry.* Translated by William Nicholson. Boston: J.T. Buckingham, 1806. Originally published as *Éléments de Chimie* (Montpellier: Jean-François Picot, 1790).

Chardin, John. *Voyages du Chevalier Chardin, en Perse, et Autres Lieux de l'Orient.* 10 vols. Paris: L. Langlès, 1811.

A Chronicle of the Carmelites in Persia and the Papal Mission of the XVIIth and XVIIIth Centuries. 2 vols. London: Eyre and Spottiswoode, 1939.

Clavijo, Ruy González de. *Embassy to Tamerlane, 1403–1406.* Translated by Guy le Strange. London: Routledge, 1928.

The Codex Mendoza. Edited by Frances Berdan and Patricia Rieff Anawalt. 4 vols. Berkeley: University of California Press, 1992.

Coste, Pascal. *Monuments Modernes de la Perse: Mesurés, Dessinés et Décrits.* Paris: A. Morel, 1867.

Curzon, George Nathaniel. *Persia and the Persian Question.* 2 vols. London: Longmans, Green, 1889.

Cushing, F.H. "Outlines of Zuni Creation Myths." In *Thirteenth Annual Report of the Bureau of Ethnology, for the Years 1891–92,* edited by J.W. Powell, 325–447. Washington DC: Smithsonian Institution, 1896.

de Boot, Boetius. *Le Parfaict Ioaillier, ou Histoire des Pierreries.* Translated by Andre Toll. Lyon: Jean Antoine Huguetan, 1644.

Department of the Interior, United States Geological Survey. *Mineral Resources of the United States.* Washington DC: Government Printing Office, 1888–92.

Description de l'Égypte ou Recuil des Observations et des Recherches Qui Ont Été Faites en Égypte Pendant l'Expedition de l'Armée Française: Histoire Naturelle. 2 vols. Paris: Imprimerie Impériale, 1813.

Dieulafait, Louis. *Diamonds and Precious Stones: A Popular Account of Gems.* New York, 1874.

Du Mans, Raphaël. *Estat de la Perse en 1660.* Edited by Charles Schefer. Paris: Ernest Leroux, 1890.

———. *Raphaël du Mans: Missionnaire en Perse au XVIIe siècle.* Edited by Francis Richard. 2 vols. Paris: L'Harmattan, 1995.

Ebers, George. *Durch Gosen zum Sinai.* Leipzig: W. Engelmann, 1872.

Ebülgazi Bahadir Han. *Histoire des Mogols et des Tartares.* Translated by Baron Desmaisons. 2 vols. Saint Petersburg: Imprimerie de l'Académie Impérial des Sciences, 1871–74.

Emanuel, Harry. *Diamonds and Precious Stones: Their History, Value, and Distinguishing Characteristics, with Simple Tests for Their Identification.* London: John Camden Hotten, 1867.

Eskandar Beg Monshi. *History of Shah 'Abbas the Great.* Translated by Roger M. Savory. 2 vols. Boulder, CO: Westview, 1978.

Evliya Çelebi. *Narrative of Travels in Europe, Asia, and Africa in the Seventeenth Century.* Translated by Joseph von Hammer. 2 vols. London: Parbury, Allen, 1834–50.

———. *An Ottoman Traveller: Selections from the Book of Travels of Evliya Çelebi.* Translated by Robert Dankoff and Sooyong Kim. London: Eland, 2010.

Exhibition of the Works of Industry of All Nations in 1851: Reports by the Jurists on the Subjects in the Thirty Classes into Which the Exhibition Was Divided. London: William Clowes and Sons, 1852.

Fadlullah Ruzbihan Khunji-Isfahani. *Persia A.D. 1478–1490: An Abridged Translation of Fadlullah b. Ruzbihan Khunji's Tarikh-i 'alam-ara-yi Amini.* Translated by Vladimir Minorsky. London: Royal Asiatic Society, 1957.

———. *Tarikh-i 'Alamara-yi Amini.* Edited by John E. Woods. Translated by Vladimir Minorsky. London: Royal Asiatic Society, 1992.

Ferrier, J.P. *Caravan Journeys and Wanderings in Persia, Afghanistan, Turkistan, and Beloochistan.* Translated by William Jesse. London: John Murray, 1856.

Feuchtwanger, L. *A Popular Treatise on Gems in Reference to Their Scientific Values.* New York: A. Hanford, 1859.

Feuvrier, Jean-Baptiste. *Trois ans à la cour de Perse.* Paris: Imprimerie Nationale, 1900.

Fischer, Gotthelf. *Essai sur le turquoise.* Moscow: Imprimerie de l'Université Impériale, 1816.

Flandin, Eugène, and Pascal Coste. *Voyage en Perse: Perse Moderne.* Vol. 8. Paris: Gide and J. Baudry, 1851.

Fraser, James Baillie. *Narrative of a Journey into Khorasan, in the Years 1821 and 1822; Including Some Account of the Countries to the North-East of Persia.* London: Longman, Hurst, Rees, Orme, Brown, and Green, 1825.

———. *Travels and Adventures in the Persian Provinces on the Southern Banks of the Caspian Sea.* London: Longman, Rees, Orme, Brown, and Green, 1826.

Fryer, John. *A New Account of East India and Persia: Being Nine Years' Travels, 1672–1681* (1698). Edited by William Crooke. 3 vols. London: Hakluyt Society, 1915.

Goethe, Johann Wolfgang von. *Goethe's Theory of Colours.* Translated by Charles Lock Eastlake. London: John Murray, 1840.

Gulbadan Begam. *The History of Humayun (Humayn-Nama).* Translated by Annette S. Beveridge. London: Royal Asiatic Society, 1902.

Hafiz-i Abru. *Jughrafiya-yi Hafiz-i Abru: Qismat-i Ruba'-i Khurasan, Herat.* Edited by Mayil Haravi. Tehran: Bunyad-i Farhangi, 1970.

———. *Zubdat al-Tavarikh.* Edited by Kamal Hajj Sayyid Javadi. 4 vols. Tehran: Sazman-i Chap va Intisharat-i Vizarat-i Farhang va Irshad-i Islami, 2001.

Hamdallah Mustawfi. *The Geographical Part of the Nuzhat-al-Qulub Composed by Hamd-Allah Mustawfi of Qazwin in 740 (1340).* Translated by Guy Le Strange. Leiden, Netherlands: E.J. Brill, 1919.

Haüy, Abbé. *Traité des Caractères Physiques de Pierres Précieuses.* Paris: Courcier, 1817.

———. *Traité de Minéralogie.* 2nd ed. 5 vols. Paris: Delance, 1822.

Hetherington, Charles. *Selim, the Nasakchi, a Persian Tale, in Verse.* London: Whittaker, 1867.

Houtum-Schindler, Albert. "Neue Angaben über die Mineralreichthümer Persiens und Notizen über die Gegerd westlich von Zendjan." In *Jahrbuch der Kaiserlich-Königlichen Geologischen Reichsanstalt*, 169–90. Vienna: K.K. Hof- und Staats-Druckerei, 1881.

———."Safarnama-yi Khurasan." In *Sih Safarnama*, ed. Qudrat Allah Rushani Za'faranlu, 145–211. Tehran: University of Tehran, 1968.

———. "The Turquoise Mines of Nishapur, Khorassan." In *Records of the Geological Survey of India*, vol. 17, 132–42. Calcutta: Geological Survey, 1884.

Hunt, Robert. *Ure's Dictionary of Arts, Manufactures, and Mines, Containing a Clear Exposition of Their Principles and Practice*. London: Longmans, Green, 1878.

Ibrahim Sahhafbashi. *Safarnama-yi Ibrahim Sahhafbashi*. Edited by Muhammad Mushiri. Tehran: Shirkat-i Mu'allifan va Mutarjiman-i Iran, 1978.

Ibn Khaldun. *The Muqaddimah: An Introduction to History*. Translated by Franz Rosenthal. 3 vols. Princeton: Princeton University Press, 1958.

'Inayat Khan. *The Shah Jahan Nama of 'Inayat Khan: An Abridged History of the Mughal Emperor Shah Jahan, Compiled by His Royal Librarian*. Edited by W.E. Begley and Z.A. Desai. Translated by A.R. Fuller. Oxford: Oxford University Press, 1990.

Iskandar Bayg Munshi. *Tarikh-i 'Alamara-yi 'Abbasi*. Edited by Iraj Afshar. 2 vols. Tehran: Amir Kabir, 2003.

I'timad al-Saltana, Muhammad Hasan Khan Sani' al-Dawla. *Matla' al-Shams, Tarikh-i Arz-i Aqdas va Mashhad-i Muqaddas, dar Tarikh va Jughrafiya-yi Mashruh-i Balad va Imakan-i Khurasan* (1882–84). 3 vols. Tehran: Farhangsara, 1986.

———. *Tarikh-i Muntazam-i Nasiri*. Edited by Muhammad Isma'il Rizvani. 3 vols. Tehran: Dunya-yi Kitab, 1963.

Jahangir. *The Jahangirnama: Memoirs of Jahangir, Emperor of India*. Translated by Wheeler M. Thackston. Oxford: Oxford University Press, 1999.

"Javahirnama." Edited by Taqi Binesh. *Farhang-i Iran Zamin* 12 (1964): 273–97.

Kaempfer, Engelbert. *Amoenitatum Exoticarum Politico-Physico-Medicarum Fasciculi V: Variae Relationes, Observationes et Descriptiones Rerum Persicarum et Ulterioris Asiae*. Lemgo, Germany: Heinrich Wilhelm Meyer, 1712.

Kamal al-Din 'Abd al-Razzaq Samarqandi. *Matla' al-Sa'dayn va Majma' al-Bahrayn*. Edited by 'Abd al-Husayn Nava'i. 2 vols. Tehran: 'Ulum-i Insani va Mutala'at-i Farhangi, 2004.

Khwaja Shams al-Din Muhammad Hafiz. *Divan-i Hafiz*. Edited by Parviz Natil Khanlari. 2 vols. Tehran: Chapkhana-yi Nil, 1980.

Khwandamir. *Habibu's-Siyar*. Vol. 3, *The Reign of the Mongol and the Turk*. Translated by Wheeler Thackston. Cambridge, MA: Harvard University Press, 1994.

Kunz, George Frederick. *Rings for the Finger: From the Earliest Known Times to the Present, with Full Descriptions of the Origin, Early Making, Materials, the Archaeology, History, for Affection, for Love, for Engagement, for Wedding, Commemorative Mourning, etc.* Philadelphia, 1917.

Laborde, Léon de, and Louis Maurice Adolphe Linant de Bellefonds. *Voyage de l'Arabie Pétrée*. Paris: Giard, 1830.

Lana Terzi, Francesco. *Podromo, Ouero, Saggio di Alcune Inventioni Nuove Premesso all'Arte Maestra*. Brescia: Rizzardi, 1670.

La Rue, François de. *De Gemmis Aliquot*. Paris: Chrétien Wechel, 1547.

Leonardus, Camillus. *The Mirror of Stones*. London: J. Freeman, 1750. Translation of *Speculum Lapidum*.

———. *Speculum Lapidum*. Venice: J.B. Sessa, 1502; reprint, Paris: Carolus Sevestre, 1610.

Mahomed Kasim Ferishta. *History of the Rise of Mahomedan Power in India*. Translated by John Briggs. Vol. 2. London: Longman, Rees, Orme, Brown, and Green, 1829.

Mandeville, John. *The Book of Marvels and Travels*. Edited by Anthony Bale. Oxford: Oxford University Press, 2012.

———. *Le Lapidaire en Francoys*. Lyon: Bernardus de Gordonio, c. 1495.

———. *Le Lapidaire en Francoys*. Lyon or Paris: Michel Le Noir, c. 1495–1500.

———. *Le Lapidaire en Francoys*. Paris: Alain Lotrian, 1544.

———. *The Travels of Sir John Mandeville*. Edited by A.W. Pollard. London: Macmillan, 1915.

Maspero, Gaston Camille Charles. *History of Egypt, Chaldea, Syria, Babylonia, and Assyria*. Translated by M.L. McClure. 4 vols. London: Grolier Society, 1903–6.

Mas'ud Mirza Zill al-Sultan. *Tarikh-i Sarguzasht-i Mas'udi*. Tehran: Intisharat-i Babak, 1983.

Matthews, Washington. *Navaho Legends*. New York: Houghton, Mifflin, 1897.

Mawe, John. *A Treatise on Diamonds and Precious Stones, Including Their History—Natural and Commercial*. London: Longman, Hurst, Rees, Orme, and Brown, 1813.

Membré, Michele. *Mission to the Lord Sophy of Persia (1539–1542)*. Translated by A.H. Morton. London: School of Oriental and African Studies, 1993.

Mirza Abu al-Hasan Khan Ilchi. *Hayratnama: Safarnama-yi Mirza Abu al-Hasan Khan Ilchi bih Landan*. Edited by Hasan Mursilvand. Tehran: Farhang-i Rasa, 1985.

Mirza I'tisam al-Din, *Shigurf Namah i Velaet, or Excellent Intelligence Concerning Europe; Being the Travels of Mirza Itesa Modeen, in Great Britain and France*. Translated by James Edward Alexander. London: Parbury, Allen, 1827.

Mohammed Ben Manssur. "Extracts from the Persian Work Called 'The Book of Precious Stones,' by Mohammed Ben Manssur." *Annals of Philosophy; or, Magazine of Chemistry, Mineralogy, Mechanics, Natural History, Agriculture and the Arts* 15 (1820): 178–90.

Mohammad Rafi' al-Din Ansari Mostowfi al-Mamalek. *Dastur al-Moluk: A Safavid State Manual*. Translated by Willem Floor and Mohammad H. Faghfoory. Costa Mesa, CA: Mazda, 2007.

Mohan Lal. *Travels in the Panjab, Afghanistan, and Turkistan, to Balk, Bokhara, and Herat*. London: Wm.H. Allen, 1846.

Muhammad al-Idrisi. *Géographie d'Édrisi.* Translated by P. A. Jaubert. Vol. 2. Paris: Imprimerie Royale, 1840.

Muhammad bin Abi Barakat Javahiri Nishapuri. *Javahirnama-yi Nizami.* Edited by Iraj Afshar and Muhammad Rasul Daryagasht. Tehran: Miras-i Maktub, 2004.

Muhammad ibn Ahmad Shams al-Din al-Muqaddasi. *Ahsan al-Taqasim fi Ma'rifat al-Aqalim.* Edited by M. J. de Goeje. Leiden, Netherlands: Brill, 1877.

Nadir Mirza. *Tarikh va Jughrafiya-yi Dar al-Saltana-yi Tabriz.* Edited by Muhammad Mushiri. Tehran: Iqbal, 1981.

Nasir al-Din Shah Qajar. *Safarnama-yi Duvvum-i Khurasan.* Tehran: Intisharat-i Kavus, 1984.

Nasir al-Din Tusi. *Tansuqnama-yi Ilkhani.* Edited by Mudarris Razavi. Tehran: Farhang-i Iran, 1969.

Newton, Isaac. *Opticks: Or, A Treatise of the Reflexions, Refractions, Inflexions and Colours of Light.* London, 1704.

Nichols, Thomas. *Lapidary; Or, The History of Precious Stones: With Cautions for the Undeceiving of All Those That Deal with Precious Stones.* Cambridge: Thomas Buck, 1652.

Orta, Garcia de. *Colloquies on the Simples and Drugs of India.* Translated by Sir Clements Markham. London: Henry Sotheran, 1913.

Ouseley, William. *Travels in Various Countries of the East; More Particularly Persia.* 3 vols. London: Rodwell and Martin, 1819.

Palmer, Edward Henry. *The Desert of the Exodus: Journeys on Foot in the Wilderness of the Forty Years' Wanderings; Undertaken in Connexion with the Ordnance Survey of Sinai and the Palestine Exploration Fund.* Vol. 2. Cambridge: Deighton, Bell, 1871.

"The Peninsula of Sinai; Notes of Travel Therein." *Littell's Living Age* 1133 (February 17, 1866): 481–502.

Réaumur, René Antoine de. "Observations sur les mines de turquoises du royaume; sur le nature de la matière qu'on y trouve, et sur la manière dont on lui donne la couleur." *Memoires de l'académie royale des sciences* (1715): 174–202.

Review of *Narrative of a Journey into Khorasan, in the Years 1821 and 1822.* In *Asiatic Journal and Monthly Miscellany, East India Company,* vol. 20 (London, 1825), 551–60.

Ringler Thomson, E. C. "The Turquoise Mines of Persia." *Windsor Magazine* 26 (1907): 61–67.

Ritter, Carl. *The Comparative Geography of Palestine and the Sinaitic Peninsula.* Vol. 1. London: J. B. Lippincott, 1865.

Roe, Sir Thomas. *The Embassy of Sir Thomas Roe to the Embassy of the Great Mogul, 1615–1619, as Narrated in His Journal and Correspondence.* 2 vols. London: Hakluyt Society, 1899.

Roero, Osvaldo. *Ricordi dei Viaggi al Cashemir, Piccolo e Medio Tibet e Turkestan.* 3 vols. Turin: Camilla E. Bertolero, 1881.

Ruznama-yi Dawlat-i 'Awliya-yi Iran. 2 vols. Tehran: National Library of the Islamic Republic of Iran, 1991.

Ruznama-yi Vaqa'i'-yi Ittifaqiya. 4 vols. Tehran: National Library of the Islamic Republic of Iran, 1994.

Sahagún, Bernardino de. *"Book Twelve of the Florentine Codex."* In *We People Here: Nahuatl Accounts of the Conquest of Mexico.* Translated and edited by James Lockhart. Berkeley: University of California Press, 1993.

———. *General History of the Things of New Spain.* Translated by Arthur J. O. Anderson and Charles Dibble. 2nd ed. 12 vols. Santa Fe: School of American Research, 1975.

Shakespeare, William. *The Excellent History of the Merchant of Venice.* London: J. Roberts, 1600.

Sharaf al-Din 'Ali Yazdi. *Zafarnama: Tarikh-i 'Umumi-yi Mufassil-i Iran dar Dawra-yi Timuriyan.* Edited by Muhammad 'Abbasi. 2 vols. Tehran: Amir Kabir, 1957.

Silliman, B. "Turquoise of New Mexico." *American Journal of Science and Arts* 122 (1881): 67–71.

Sowerby, James. *Exotic Mineralogy: Or, Coloured Figures of Foreign Minerals as a Supplement to British Mineralogy.* 2 vols. London: Benjamin Meredith, 1811–17.

Tadhkirat al-Muluk. Edited by Muhammad Dabirsiyaqi. Tehran: Amir Kabir, 1989.

Tadhkirat al-Muluk: A Manual of Safavid Administration (circa 1137/1725). Translated by V. Minorsky. Cambridge: Cambridge University Press, 1943.

Tarikh al-Hind al-Gharbi al-Musamma bi-Hadith-i Naw. Istanbul: Ibrahim Muteferrika, 1729.

Tavernier, Jean Baptiste. *Collections of Travels through Turkey into Persia, and the East Indies.* 3 vols. London: Moses Pitt, 1684.

———. *Travels in India: Translated from the Original French Edition of 1676.* Translated by V. Ball. London: Macmillan, 1889.

Tifaschi, Ahmad ibn Yusuf al-. *Arab Roots of Gemology: Ahmad ibn Yusuf al-Tifaschi's "Best Thoughts on the Best of Stones."* Translated by Samar Najm Abul Huda. Lanham, MD: Scarecrow, 1998.

Wilson, Charles W. *Picturesque Palestine: Sinai and Egypt.* Vol. 4. London: J. S. Virtue, 1881.

Yate, C. E. *Northern Afghanistan: Or Letters from the Afghan Boundary Commission.* Edinburgh: William Blackwood and Sons, 1888.

Zahir al-Din Babur. *The Baburnama: Memoirs of Babur, Prince and Emperor.* Translated by Wheeler M. Thackston. New York: Random House, 2002.

SECONDARY MATERIALS

Afshar, Iraj. "Javaher-name-ye Nezami." In *Nasir al-Din Tusi, philosophe et savant de XIIIe siècle,* edited by Z. Vesel, Afshar, and Parviz Mohebbi, 151–165. Tehran: Presses Universitaires d'Iran, 2000.

Akasoy, Anna, Charles Burnett, and Ronit Yoeli-Tlalim, eds. *Islam and Tibet: Interactions along the Musk Routes.* London: Ashgate, 2010.

Alam, Muzaffar, and Sanjay Subrahmanyam. *Indo-Persian Travels in the Age of Discoveries, 1400–1800.* Cambridge: Cambridge University Press, 2007.

Allan, James W. "Abu'l-Qasim's Treatise on Ceramics." *Iran* 11 (1973): 111–20.

———. "Early Safavid Metalwork." In *Hunt for Paradise: Court Arts of Safavid Iran, 1501–1576*, edited by Jon Thompson and Sheila R. Canby, 203–40. Milan: Skira, 2003.

Allsen, Thomas. *Commodity and Exchange in the Mongol Empire: A Cultural History of Islamic Textiles.* Cambridge: Cambridge University Press, 1997.

———. "Mongolian Princes and Their Merchant Partners, 1200–1260." *Asia Minor* 2, no. 2 (1989): 83–125.

Amanat, Abbas. *Cities and Trade: Consul Abbott on the Economy and Society of Iran, 1847–1866.* London: Ithaca, 1983.

———. *Pivot of the Universe: Nasir al-Din Shah and the Iranian Monarchy, 1831–1896.* Berkeley: University of California Press, 1997.

Ambraseys, N. N., and Charles Melville. *A History of Persian Earthquakes.* Cambridge: Cambridge University Press, 1982.

Anderson, Frank J. *Riches of the Earth: Ornamental, Precious, and Semiprecious Stones.* New York: Rutledge, 1981.

Appadurai, Arjun, ed. *The Social Life of Things: Commodities in Cultural Perspective.* Cambridge: Cambridge University Press, 1986.

Asher, Catherine. *Architecture of Mughal India.* Cambridge: Cambridge University Press, 1992.

Aslanian, Sebouh. *From the Indian Ocean to the Mediterranean: The Global Trade Networks of Armenian Merchants from New Julfa.* Berkeley: University of California Press, 2011.

Aube, Sandra. "La Mosquée bleue de Tabriz (1465): Remarques sur la céramique architecturale Qara Qoyunlu." *Studia Iranica* 37, no. 2 (2008): 241–77.

Aubin, Jean. "Réseau pastoral et réseau caravanier: Les grand'routes du Khurassan à l'époque Mongole." *Le Monde Iranien et l'Islam* 1 (1971): 105–30.

Avery, Peter, Gavin Hambly, and Charles Melville, eds. *From Nadir Shah to the Islamic Republic.* Vol. 7 of *The Cambridge History of Iran.* Cambridge: Cambridge University Press, 1991.

Babaie, Sussan. *Isfahan and Its Palaces: Statecraft, Shi'ism, and the Architecture of Conviviality in Early Modern Iran.* Edinburgh: Edinburgh University Press, 2008.

Balfour-Paul, Jenny. *Indigo.* London: British Museum Press, 1998.

Ball, Philip. *Bright Earth: The Invention of Colour.* Chicago: University of Chicago Press, 2001.

Bang, Peter Fibiger, and C. A. Bayly, eds. *Tributary Empires in Global History.* Basingstoke: Palgrave Macmillan, 2011.

Barkey, Karen. *Empire of Difference: The Ottomans in Comparative Perspective.* Cambridge: Cambridge University Press, 2008.

Barry, Michael. *Design and Color in Islamic Architecture: Eight Centuries of the Tile Maker's Art.* New York, 1996.

Bastani Parizi, Muhammad Ibrahim. *Siyasat va Iqtisad-i 'Asr-i Safavi.* Tehran: Intisharat-i Safi 'Ali Shah, 1999.

Baxandall, Michael. *Painting and Experience in Fifteenth-Century Italy.* Oxford: Oxford University Press, 1972.

Bayly, C. A. *Empire and Information*. Cambridge: Cambridge University Press, 1996.

Beresneva, L. *The Decorative and Applied Art of Turkmenia*. Leningrad: Aurora, 1976.

Berrie, Barbara H. "Pigments in Venetian and Islamic Painting." In *Venice and the Islamic World, 828–1797*, edited by Stefano Carboni, 140–45. New Haven: Yale University Press, 2007.

Blair, Sheila S., and Jonathan Bloom. *And Diverse Are Their Hues: Color in Islamic Art and Culture*. New Haven: Yale University Press, 2011.

———. *The Art and Architecture of Islam, 1250–1800*. New Haven: Yale University Press, 1994.

Blake, Stephen. *Half of the World: The Social Architecture of Safavid Isfahan*. Costa Mesa, CA: Mazda, 1999.

———. *Shahjahanabad: The Sovereign City in Mughal India, 1639–1739*. Cambridge: Cambridge University Press, 2002.

Bleichmar, Daniela. *Visible Empire: Botanical Expeditions and Visual Culture in the Hispanic Enlightenment*. Chicago: University of Chicago Press, 2012.

Blunt, Wilfrid. *Isfahan: Pearl of Persia*. 1966; reprint, London, 2009.

Bosworth, C. E. "The Early Islamic History of Ghur." *Central Asiatic Journal* 6 (1961): 116–33.

Bourriau, Janine. "The Second Intermediate Period (*c.*1650–1550 BC)." In *The Oxford History of Ancient Egypt*, edited by Ian Shaw, 172–206. Oxford: Oxford University Press, 2000.

Braudel, Fernand. *The Mediterranean and the Mediterranean World in the Age of Philip II*. Translated by Sian Reynolds. 2 vols. Berkeley: University of California Press, 1995.

Breasted, James Henry. *A History of Egypt, from the Earliest Times to the Persian Conquest*. New York: Charles Scribner's Sons, 1905.

Brook, Timothy. *Vermeer's Hat: The Seventeenth Century and the Dawn of the Global World*. New York: Bloomsbury, 2008.

Browne, Edward G. "The Persian Manuscripts of the Late Albert Houtum-Schindler." *Journal of the Royal Asiatic Society of Great Britain and Ireland* 49 (October 1917): 657–94.

Brummett, Palmira. *Ottoman Seapower and Levantine Diplomacy in the Age of Discovery*. Albany, New York: State University of New York Press, 1994.

Bulliett, Richard. *Cotton, Climate, and Camels in Early Islamic Iran: A Moment in World History*. New York: Columbia University Press, 2009.

———. *The Patricians of Nishapur: A Study in Medieval Islamic Social History*. Cambridge, MA: Harvard University Press, 1972.

Burke, Edmund, III. "Islamic History as World History: Marshall Hodgson, 'The Venture of Islam.'" *International Journal of Middle East Studies* 10, no. 2 (1979): 241–64.

———. "Marshall G. S. Hodgson and the Hemispheric Interregional Approach to World History." *Journal of World History* 6, no. 2 (1995): 237–50.

Burke, Edmund, III, and Ken Pomeranz, eds. *The Environment and World History*. Berkeley: University of California Press, 2009.

Callender, Gae. "The Middle Kingdom Renaissance (c.2055–1650 BC)." In *The Oxford History of Ancient Egypt,* edited by Ian Shaw, 137–71. Oxford: Oxford University Press, 2000.

Canby, Sheila. *Shah 'Abbas: The Remaking of Iran.* London: British Museum Press, 2009.

Canizares-Esguerra, Jorge. *Nature, Empire, and Nation: Explorations in the History of Science in the Iberian World.* Stanford, CA: Stanford University Press, 2006.

Carboni, Stefano, ed. *Venice and the Islamic World, 828–1797.* New Haven: Yale University Press, 2007.

Casale, Giancarlo. *The Ottoman Age of Exploration.* Oxford: Oxford University Press, 2010.

Çelik, Zeynep. *Empire, Architecture, and the City: French-Ottoman Encounters, 1830–1914.* Seattle: University of Washington Press, 2008.

Çelik, Zeynep, Julia Clancy-Smith, and Francis Terpak, eds. *Walls of Algiers: Narratives of the City through Text and Image.* Seattle: University of Washington Press, 2009.

Chenciner, Robert. *Madder Red, a History of Luxury and Trade, Plant Dyes and Pigments in World Commerce and Art.* London: Routledge, 2000.

Christian, David. "Silk Roads or Steppe Roads? The Silk Roads in World History." *Journal of World History* 11, no. 1 (2000): 1–26.

Cook, Harold J. *Matters of Exchange: Commerce, Medicine, and Science in the Dutch Golden Age.* New Haven: Yale University Press, 2007.

Cooney, John. "Major Macdonald, a Victorian Romantic." *Journal of Egyptian Archaeology* 58 (August 1972): 280–85.

Crosby, Alfred. *Ecological Imperialism: The Biological Expansion of Europe, 900–1900.* Cambridge: Cambridge University Press, 2004.

Dalby, Andrew. *Dangerous Tastes: The Story of Spices.* Berkeley: University of California Press, 2000.

Dale, Stephen. *Indian Merchants and Eurasian Trade, 1600–1750.* Cambridge: Cambridge University Press, 1994.

———. *The Muslim Empires of the Ottomans, Safavids, and Mughals.* Cambridge: Cambridge University Press, 2010.

Dani, Ahmad Hasan. *Thatta: Islamic Architecture.* Islamabad: Institute of Islamic History, Culture, and Civilization, 1982.

Davis, Diana. *Resurrecting the Granary of Rome: Environmental History and French Colonial Expansion in North Africa.* Ohio: University of Ohio Press, 2008.

Davis, Diana, and Edmund Burke III. *Environmental Imaginaries of the Middle East and North Africa.* Ohio: University of Ohio Press, 2012.

Dickson, Martin B. "The Fall of the Safavi Dynasty." *Journal of the American Oriental Society* 82, no. 4 (1962): 503–17.

Di Cosmo, Nicola, Allen Frank, and Peter Golden, eds. *The Cambridge History of Inner Asia: The Chinggisid Age.* Cambridge: Cambridge University Press, 2009.

Desautels, Paul E. *The Mineral Kingdom.* New York: Madison Square, 1968.

Eaton, Richard. *A Social History of the Deccan, 1300–1761: Eight Indian Lives.* Cambridge: Cambridge University Press, 2005.

Eckenstein, Lina. *A History of Sinai.* New York: Macmillan, 1921.

Finlay, Robert. *The Pilgrim Art: Cultures of Porcelain in World History.* Berkeley: University of California Press, 2010.

———. "Weaving the Rainbow: Visions of Color in World History." *Journal of World History* 18, no. 4 (2007): 383–431.

Finlay, Victoria. *Color: A Natural History of the Palette.* New York: Random House, 2002.

Fitzhugh, Elisabeth West, and Willem Floor. "Cobalt." *Encylcopaedia Iranica* 5, no. 8 (1992): 873–75.

Fletcher, Joseph. "Integrative History: Parallels and Interconnections in the Early Modern Period, 1500–1800." In *Studies on Chinese and Islamic Inner Asia,* edited by Beatrice Forbes Manz, 1–35. London: Variorum, 1995.

Flood, Finbarr B. *Objects of Translation: Material Culture and Medieval "Hindu-Muslim" Encounter.* Princeton: Princeton University Press, 2009.

Floor, Willem. *The Economy of Safavid Persia.* Wiesbaden: Ludwig Reichert, 2000.

———. *The Persian Textile Industry in Historical Perspective, 1500–1925.* Paris: L'Harmattan, 1999.

Floor, Willem, and Edmund Herzig, eds. *Iran and the World in the Safavid Age.* London: I. B. Tauris, 2012.

Freedman, Paul. *Out of the East: Spices and the Medieval Imagination.* New Haven: Yale University Press, 2008.

Frye, Richard. "Firuzkuh." In *Encyclopedia of Islam.* 2nd ed. Vol. 2, 928. Leiden, Netherlands: E. J. Brill, 1965.

Gardiner, Sir Alan. *Egypt of the Pharaohs: An Introduction.* Oxford: Oxford University Press, 1961.

Gilbar, Gad. "The Opening Up of Qajar Iran: Some Economic and Social Aspects." *Bulletin of the School of Oriental and African Studies* 49, no. 1 (1986): 76–89.

Girayili, Firaydun. *Nishabur: Shahr-i Firuza.* Mashhad: Firdawsi University, 1978.

Golombek, Lisa, and Donald Wilber. *The Timurid Architecture of Iran and Turan.* 2 vols. Princeton: Princeton University Press, 1988.

Green, Nile. *Bombay Islam: The Religious Economy of the West Indian Ocean, 1840–1915.* Cambridge: Cambridge University Press, 2011.

Green, Nile, and James Gelvin, eds. *Global Muslims in the Age of Steam and Print.* Berkeley: University of California Press, 2014.

Greenblatt, Stephen. *Marvelous Possessions: The Wonder of the New World.* Chicago: University of Chicago Press, 1991.

Grimal, Nicolas. *A History of Ancient Egypt.* Oxford: Oxford University Press, 1988.

Grove, Richard. *Green Imperialism: Colonial Expansion, Tropical Island Edens and the Origins of Environmentalism, 1600–1860.* Cambridge: Cambridge University Press, 1995.

Guli, Amin. *Tarikh-i Siyasi va Ijtima'i-yi Turkmanha*. Tehran: Nashr-i 'Ilm, 1987.

Gurney, John. "Albert Houtum-Schindler." *Encyclopaedia Iranica* 12, no. 5 (2003): 540–43.

Haider, Najaf. "Precious Metal Flows and Currency Circulation in the Mughal Empire." In "Money in the Orient," special issue, *Journal of the Economic and Social History of the Orient* 39, no. 3, (1996): 298–364.

Hamadeh, Shirine. *The City's Pleasures: Istanbul in the Eighteenth Century*. Seattle: University of Washington Press, 2007.

Hodge, F. W. *Turquois Work of the Hawikuh, New Mexico*. New York: Museum of the American Indian, 1921.

Hodgson, Marshall. "The Interrelations of Societies in History." *Comparative Studies in Society and History* 5, no. 2 (1963): 227–50.

———. "The Role of Islam in World History." *International Journal of Middle East Studies* 1, no. 2 (1970): 99–123.

———. *The Venture of Islam: Conscience and History in a World Civilization*. 3 vols. Chicago: University of Chicago Press, 1974.

Iqbal, Abbas. "Tarikh-i Javahir dar Iran." *Farhang-i Iran Zamin* 9 (1961): 5–45.

Islam, Riazul. *Indo-Persian Relations: A Study of the Political and Diplomatic Relations between the Mughul Empire and Iran*. Tehran: Intisharat-i Bunyad-i Farhang-i Iran, 1970.

Issawi, Charles. *The Economic History of Iran, 1800–1914*. Chicago: University of Chicago Press, 1971.

Jackson, Peter, and Laurence Lockhart, eds. *The Timurid and Safavid Periods*. Vol. 6 of *The Cambridge History of Iran*. Cambridge: Cambridge University Press, 1986.

Jardine, Lisa. *Worldly Goods: A New History of the Renaissance*. New York: Doubleday, 1996.

Jones, Fayette Alexander. *New Mexico Mines and Minerals*. Santa Fe: New Mexican Printing Company, 1904.

Kashani-Sabet, Firoozeh. *Frontier Fictions: Shaping the Iranian Nation, 1804–1946*. Princeton: Princeton University Press, 1999.

Keay, John. *The Spice Route: A History*. Berkeley: University of California Press, 2006.

Keddie, Nikki. "Material Culture and Geography: Toward a Holistic Comparative History of the Middle East." *Comparative Studies in Society and History* 26, no. 4 (October 1984): 709–34.

Kläy, Ernst Johannes, and Anne Brechbühl. *Islamic Collection Henri Moser Charlottenfels*. Bern: Musée d'histoire de Berne, 1991.

Komaroff, Linda. *Gifts of the Sultan: The Arts of Giving at the Islamic Courts*. New Haven: Yale University Press, 2011.

———. *The Gift Tradition in Islamic Art*. New Haven: Yale University Press, 2012.

Koseoglu, Cengiz. *The Topkapi Saray Museum: The Treasury*. Translated and edited by J. M. Rogers. Boston: Little, Brown, 1987.

Lambton, Ann. "'Piskash': Present or Tribute?" *Bulletin of the School of Oriental and African Studies* 57 (1994): 145–58.

Lane, Kris. *Colour of Paradise: The Emerald in the Age of Gunpowder Empires.* New Haven: Yale University Press, 2010.

Laufer, Berthold. *Notes on Turquois in the East.* Chicago: Field Museum of Natural History, 1913.

Lockhart, Laurence. *The Fall of the Safavi Dynasty and the Afghan Occupation of Persia.* Cambridge: Cambridge University Press, 1958.

Mack, Rosamond. *Bazaar to Piazza: Islamic Trade and Italian Art, 1300–1600.* Berkeley: University of California Press, 2002.

Malek, Jaromir. "The Old Kingdom (*c.*2686–2160 BC)." In *The Oxford History of Ancient Egypt,* edited by Ian Shaw, 83–107. Oxford: Oxford University Press, 2000.

Manniche, Lise. *Sacred Luxuries: Fragrance, Aromatherapy, and Cosmetics in Ancient Egypt.* Ithaca: Cornell University Press, 1999.

Manz, Beatrice. *Power, Politics, and Religion in Timurid Iran.* Cambridge: Cambridge University Press, 2007.

———. *The Rise and Rule of Tamerlane.* Cambridge: Cambridge University Press, 1999.

Matthee, Rudolph P. (Rudi). *Persia in Crisis: Safavid Decline and the Fall of Isfahan.* London: I.B. Tauris, 2012.

———. *The Politics of Trade in Safavid Iran: Silk for Silver, 1600–1730.* Cambridge: Cambridge University Press, 1999.

———. *The Pursuit of Pleasure: Drugs and Stimulants in Iranian History, 1500–1900.* Princeton: Princeton University Press, 2005.

———. "The Safavids under Western Eyes: Seventeenth-Century European Travelers to Iran." *Journal of Early Modern History* 13 (2009): 137–71.

Mauss, Marcel. *The Gift: The Form and Reason for Exchange in Archaic Societies.* Translation of *Essai sur le Don: Forme et raison de l'échange dans les sociétés archaïques* (Paris: Presses Universitaires, 1950) by W.D. Walls. London: Routledge, 1990.

McChesney, Robert. "'Barrier of Heterodoxy'?: Rethinking the Ties between Iran and Central Asia in the Seventeenth Century." In *Safavid Persia: The History and Politics of an Islamic Society,* ed. Charles Melville, 231–67. London: I.B. Tauris, 1996.

McEwan, Colin, Andrew Middleton, Caroline Cartwright, and Rebecca Stacey. *Turquoise Mosaics from Mexico.* Durham: Duke University Press, 2006.

Melloy, Ellen. *The Anthropology of Turquoise: Reflections on Desert, Sea, Stone, and Sky.* New York: Vintage, 2002.

Melville, Charles. "Earthquakes in the History of Nishapur." *Iran* 18 (1980): 103–20.

———. "Historical Monuments and Earthquakes in Tabriz." *Iran* 29 (1981): 159–77.

———, ed. *Safavid Persia: The History and Politics of an Islamic Society.* London: I.B. Tauris, 1996.

Metcalf, Thomas. *Ideologies of the Raj.* Berkeley: University of California Press, 1995.

———. *An Imperial Vision: Indian Architecture and Britain's Raj.* Berkeley: University of California Press, 1989.

Mikhail, Alan. *Nature and Empire in Ottoman Egypt.* Cambridge: Cambridge University Press, 2011.

——, ed. *Water on Sand: Environmental Histories of the Middle East and North Africa.* Oxford: Oxford University Press, 2012.

Mintz, Sidney. *Sweetness and Power: The Place of Sugar in Modern History.* New York: Viking, 1985.

Mitchell, George, and Richard Eaton. *Firuzabad: Palace City of the Deccan.* Oxford: Oxford University Press, 1992.

Mitchell, Timothy. *Colonising Egypt.* Berkeley: University of California Press, 1991.

Moin, A. Azfar. *The Millennial Sovereign: Sacred Kingship and Sainthood in Islam.* New York: Columbia University Press, 2012.

Mottahedeh, Roy. " '*Aja'ib* in *The Thousand and One Nights.*" In *"The Thousand and One Nights" in Arabic Literature and Society,* edited by Richard Hovanissian and Georges Sabagh, 29–39. Cambridge: Cambridge University Press, 1997.

Necipoglu, Gulru. *The Age of Sinan: Architectural Culture in the Ottoman Empire.* Princeton: Princeton University Press, 2005.

——. "Framing the Gaze in Ottoman, Safavid, and Mughal Palaces." *Ars Orientalis* 23 (1993): 303–42.

Neese, William D. *Introduction to Mineralogy.* Oxford: Oxford University Press, 2000.

Okihiro, Gary. *Pineapple Culture: A History of the Tropical and Temperate Zones.* Berkeley: University of California Press, 2009.

Orfali, Bilal. "The Works of Abu Mansur al-Tha'alibi (350–429/961–1039)." *Journal of Arabic Literature* 40 (2009): 273–318.

Pamuk, Şevket. *A Monetary History of the Ottoman Empire.* Cambridge: Cambridge University Press, 2000.

Pastoureau, Michel. *Blue: The History of a Color.* Princeton: Princeton University Press, 2001.

Patel, Alka. *Building Communities in Gujarat: Architecture and Society during the Twelfth through Fourteenth Centuries.* Leiden, Netherlands: Brill, 2004.

Petrie, W.M. Flinders. *Researches in Sinai.* London: E.P. Dutton, 1906.

Pogue, Joseph. *The Turquois: A Study of Its History, Mineralogy, Geology, Ethnology, Archaeology, Mythology, Folklore and Technology.* Washington DC: National Academy of Sciences, 1915.

Pomeranz, Ken. *The Great Divergence: China, Europe, and the Making of the World Economy.* Princeton: Princeton University Press, 2000.

Pratt, Mary Louise. *Imperial Eyes: Travel Writing and Transculturation.* London: Routledge, 1992.

Raj, Kapil. *Relocating Modern Science: Circulation and the Construction of Knowledge in South Asia and Europe.* Basingstoke: Palgrave Macmillan, 2007.

Richards, John. *The Imperial Monetary System of Mughal India.* Delhi: Oxford University Press, 1987.

——. *The Mughal Empire.* Cambridge: Cambridge University Press, 1996.

————. *The Unending Frontier: An Environmental History of the Early Modern World*. Berkeley: University of California Press, 2003.

Rizvi, Kishwar. *The Safavid Dynastic Shrine: Architecture, Religion, and Power in Early Modern Iran*. London: Tauris Academic Studies, 2011.

Roemer, H.R. "The Safavid Period." In *The Timurid and Safavid Periods*, vol. 6 of *The Cambridge History of Iran*, eds. Peter Jackson and Laurence Lockhart, 189–350. Cambridge: Cambridge University Press, 1986.

Said, Edward. *Culture and Imperialism*. New York: Vintage, 1994.

————. *Orientalism*. New York: Vintage, 1978.

Sajdi, Dana, ed. *Ottoman Tulips, Ottoman Coffee: Leisure and Lifestyle in the Eighteenth Century*. London: Tauris Academic Studies, 2008.

Schiebinger, Londa. *Plants and Empire: Colonial Bioprospecting in the Atlantic World*. Cambridge, MA: Harvard University Press, 2007.

Schiebinger, Londa, and Claudia Swan, eds. *Colonial Botany: Science, Commerce, and Politics in the Early Modern World*. Philadelphia: University of Pennsylvania Press, 2007.

Shaw, Ian. "Egypt and the Outside World." In *The Oxford History of Ancient Egypt*, edited by Shaw, 308–23. Oxford: Oxford University Press, 2000.

————, ed. *The Oxford History of Ancient Egypt*. Oxford: Oxford University Press, 2000.

Smith, Pamela, and Paula Findlen, eds. *Merchants and Marvels: Commerce, Science, and Art in Early Modern Europe*. London, 2002.

Smith, Pamela, and Benjamin Schmidt, eds. *Making Knowledge in Early Modern Europe: Practices, Objects, and Texts, 1400–1800*. Chicago: University of Chicago Press, 2007.

Sohrabi, Naghmeh. *Taken for Wonder: Nineteenth-Century Travel Accounts from Iran to Europe*. Oxford: Oxford University Press, 2012.

Stein, Sara. *Plumes: Ostrich Feathers, Jews, and a Lost World of Global Commerce*. New Haven: Yale University Press, 2008.

Storey, C.A. *Persian Literature: A Bibliographic Survey*. Vol. 2, pt. 3. Leiden, Netherlands: Brill, 1977.

Subrahmanyam, Sanjay. "Connected Histories: Notes towards a Reconfiguration of Early Modern Eurasia." *Modern Asian Studies* 31, no. 3 (1997): 735–62.

————. *Explorations in Connected History: From the Tagus to the Ganges*. Oxford: Oxford University Press, 2005.

————. *Explorations in Connected History: Mughals and Franks*. Oxford: Oxford University Press, 2005.

————. "Precious Metal Flows and Prices in Western and Southern Asia, 1500–1750." *Studies in History* 7, no. 1 (1991): 79–105.

Subtelny, Maria. *Timurids in Transition: Turko-Persian Politics and Acculturation in Medieval Iran*. Leiden, Netherlands: Brill, 2007.

Tabataba'i, Sayyid Jamal Turabi. *Naqshha va Nigashtaha-yi Masjid-i Kabud-i Tabriz*. Tabriz: Shafaq-i Tabriz, 1969.

Tavakoli-Targhi, Mohamad. *Refashioning Iran: Orientalism, Occidentalism, and Historiography*. Basingstoke: Palgrave Macmillan, 2001.

Thomas, Nicholas. *Entangled Objects: Exchange, Material Culture, and Colonialism in the Pacific*. Cambridge, MA: Harvard University Press, 1991.

Thompson, Jon, and Sheila R. Canby, eds. *Hunt for Paradise: Court Arts of Safavid Iran, 1501–1576.* Milan: Skira, 2003.

Tosi, M. "The Problem of Turquoise in the Protohistoric Trade on the Iranian Plateau." *Studi di Paletnologia, Paleoantropologia, Paleontologia e Geologia del Quaternario* 2 (1974): 147–62.

Turner, Jack. *Spice: The History of a Temptation.* New York: Alfred A. Knopf, 2004.

van Dijk, Jacobus. "The Amarna Period and the Later New Kingdom (*c.*1352–1069 BC)." In *The Oxford History of Ancient Egypt,* edited by Ian Shaw, 265–307. Oxford: Oxford University Press, 2000.

Vogelsang, Willem. *The Afghans.* London: Wiley-Blackwell, 2001.

Wakefield, André. *The Disordered Police State: German Cameralism as Science and Practice.* Chicago: University of Chicago Press, 2009.

———. "Leibniz in the Mines." *Osiris* 25 (2010): 171–88.

Weigand, Phil, and Acelia Garcia de Weigand. "A Macroeconomic Study of the Relationship between the Ancient Cultures of the American Southwest and Mesoamerica." In *The Roads to Aztlan: Art from a Mythic Homeland,* edited by Virginia Fields and Victor Zamudio-Taylor, 184–95. Los Angeles: Los Angeles County Museum of Art, 2001.

Welland, Michael. *Sand: The Never-Ending Story.* Berkeley: University of California Press, 2009.

White, Sam. *The Climate of Rebellion in the Early Modern Ottoman Empire.* Cambridge: Cambridge University Press, 2011.

Woods, John E. *The Aqquyunlu: Clan, Confederation, and Empire.* Salt Lake City: University of Utah Press, 1999.

Index

TEXT
10/13 Sabon

DISPLAY
Sabon

COMPOSITOR
IDS Infotech, Ltd.

CARTOGRAPHER
Gillian Schwartz

PRINTER AND BINDER
Maple Press